Handbook of Pollution Prevention and Cleaner Production Vol. 4

Handbook of Pollution Prevention and Cleaner Production Vol. 4

Clean Electricity through Advanced Coal Technologies

Nicholas P. Cheremisinoff

ELSEVIER

AMSTERDAM • BOSTON • HEIDELBERG • LONDON • NEW YORK • OXFORD
PARIS • SAN DIEGO • SAN FRANCISCO • SINGAPORE • SYDNEY • TOKYO
William Andrew is an imprint of Elsevier

William Andrew is an imprint of Elsevier
The Boulevard, Langford Lane, Kidlington, Oxford, OX5 1GB, UK
225 Wyman Street, Waltham, MA 02451, USA

First published 2012

British Library Cataloguing in Publication Data
A catalogue record for this book is available from the British Library

Library of Congress Control Number: 2012936055

ISBN: 978-1-43-777815-1

For information on all William Andrew publications
visit our website at **store.elsevier.com**

Printed and bound in the United States

12 13 14 15   10 9 8 7 6 5 4 3 2 1

Contents

Preface

The quality of life throughout most of the world is tied to electricity. In this regard, coal plays a vital role in electricity generation worldwide. Coal-fired power plants currently fuel 41% of global electricity. In some countries, coal fuels a higher percentage of electricity. Compared to other electricity-generating technologies, it is and is likely to remain the most economical fuel source through the next generation.

Coal also carries with it the largest environmental footprint of all known power-producing technologies. Coal-fired power plants are the largest industry source of greenhouse gases. And the solid waste issues that are the result of depending on this technology are almost mind boggling. In the United States alone, roughly 1.05 billion tons of coal are burned each year to generate electricity. More than 500 coal-burning plants across the United States emit 1.9 billion tons of carbon dioxide each year and create 120 million tons of toxic waste. Coal combustion waste (CCW) constitutes the second largest waste stream after municipal solid waste in the U.S.

Sulfur dioxide scrubbers, which are used for air pollution controls, also create toxic waste. The flue-gas desulfurization (FGD) process creates a wet solid residue containing calcium sulfite ($CaSO_3$) and calcium sulfate ($CaSO_4$). Dry material such as fly ash is added to stabilize the sludge for transport and landfill storage.

Coal waste is disposed of in landfills generally referred to as ponds or surface impoundments. The ponds are lined with compacted clay soil, a plastic sheet, or both. When rain filters through the toxic ash pits, the toxic metals are leached out and pushed downward by gravity toward the lining and the soil below. Because all liners eventually degrade, crack, or tear, landfills eventually leak and release their toxins into the groundwater. The Environmental Protection Agency has reported that a 10-acre landfill would leak 0.2 to 10 gallons per day, or between 730 and 36,500 gallons over a 10-year period. There are an estimated 156 ash waste ponds across the United States containing more than 19 million tons of toxic ash waste. In 2008 one of these waste ponds owned by the Tennessee Valley Authority failed, releasing 5.4 million cubic yards (4 million m^3) into a community.

Coal has been dubbed by much of the general public as "dirty coal." It is the cause of major air pollution, global warming, and has resulted in the release of toxic heavy metals and radionuclides, which place communities at long-term health risks. The coal industry was exempt from reporting its air emissions to the public until about the turn of the twenty-first century.

This volume serves as primer on coal-burning technologies for electricity generation. Newer generation technologies that are emerging and being commercialized have significantly smaller environmental footprints than conventional technologies used with the existing coal-fired infrastructure that currently exists throughout much of the world. This volume is a part of the Cleaner Production series and meant to provide basic information and awareness of environmental issues and responsible pollution management.

The author extends a heartfelt gratitude to the publisher for its fine production of this volume and series.

Nicholas P. Cheremisinoff, Ph.D.

About the Author

Nicholas P. Cheremisinoff has 30 years of international consulting and project management experience across a broad spectrum of industry sectors dealing with environmental management practices and technologies. His assignments have focused on corporate responsible care through the application of pollution prevention and cleaner production. He has served as both a practitioner and an educator to industry stakeholders on responsible care practices that strive to improve environmental performance through the application of cost-effective clean technologies. He has designed and implemented projects for the United States Agency for International Development, the World Bank Organization, the European Union, the U.S. Trade and Development Agency, the U.S. Department of Energy, the U.S. Department of Defense, and a number of Fortune 100 and multinational companies. He has applied his experience and knowledge in major environmental toxic tort litigations as an expert witness in the fields of environmental forensics, in historical reconstruction of environmental management practices, and in formulating opinions regarding levels of responsible care exercised by industry stakeholders. He is the author, coauthor, or editor of more than 160 book publications and numerous state-of-the-art review articles and R&D papers. He received his B.S., M.S., and Ph.D. degrees in chemical engineering from Clarkson College of Technology.

1 Air Pollution

Chapter Outline

1.1 Overview

Nearly 50% of the electricity generated in the world uses coal as the primary fuel. The function of the coal-fired thermal power plant is to convert the energy available in the coal to electricity. Coal power plants work by using several steps to convert stored energy in coal to usable electricity. The conversion from coal to electricity takes place in three stages.

Stage 1 involves the conversion of energy in a boiler. Coal is burned in the boiler furnace to produce heat. Carbon in the coal and oxygen in the air combine to produce carbon dioxide and heat.

Stage 2 is a thermodynamic process. The heat from combustion of the coal boils water in the boiler to produce steam. In a modern power plant, boilers produce steam at high pressures and temperatures. The steam is then piped to a turbine. The high-pressure steam impinges and expands across a number of sets of blades in the turbine. The impulse and the thrust created rotate the turbine. The steam is then condensed and pumped back into the boiler to repeat the cycle.

In **Stage 3** rotation of the turbine rotates the generator rotor to produce electricity based on Faraday's Principle of electromagnetic induction.

Handbook of Pollution Prevention and Cleaner Production Vol. 4. DOI: http://dx.doi.org/10.1016/B978-1-4377-7815-1.00001-1

In practice, to effect these three stages of conversion, many systems and subsystems have to be in service. Also involved are different technologies, such as combustion, aerodynamics, heat transfer, thermodynamics, pollution control, and logistics. As an example, consider the following: for a coal-fired power plant of capacity 500 MW, about 2 million tons of coal are needed each year to produce the continuous power. Also, coal combustion in the boiler requires roughly 1.6 million cubic meters (m^3) of air in an hour to be delivered by fans into the furnace. For this same size plant, the ash generated from the combustion of coal is around 200,000 tons per year. Electrostatic precipitators (ESPs) capture most of this ash without dispersing it to the atmosphere. Pollutants from coal power plants include carbon dioxide, sulfur dioxide, nitrogen oxide, and particulate matter (PM). Thermal power plants are the largest producers of carbon dioxide.

The boiler for typical 500 megawatt (MW) units produces around 1,600 tons per hour of steam at a temperature of 540–600°C. The steam pressure is in the range of 200 bar. The boiler materials are designed to withstand these conditions with special consideration for operational safety. Heat transfer from the hot combustion gases to the water in the boiler takes place due to radiation and convection. The electrical generators carry large electric currents that produce heat and are cooled by hydrogen and water. The steam leaving the turbine is condensed, and the water is pumped back for reuse in the boiler. To condense all the steam requires around 50,000 cubic meters per hour of cooling water to be circulated from lakes, rivers, or the sea. The water is returned to the source with only an increase of 3–4°C to minimize negative impacts to the environment. In addition to the cooling water, the power plant also requires around 400 cubic meters per day of fresh water for making up the losses in the water–steam cycle.

Electricity has often been characterized as the foundation of advanced countries and societies. It unquestionably is the basis by which societies maintain human life and sustain and grow economies. But the dependence on coal-fired technologies is a dirty business that has global implications on long-term sustainability. In the United States alone, there are more than 440 power plants larger than 25 MWs located in 46 states and Puerto Rico that burn coal to generate electricity. The National Emissions Inventory shows that 84 of the 187 hazardous air pollutants (HAPs) identified by the U.S. Environmental Protection Agency (EPA) pose a threat to human health and the environment. These plants release about 386,000 tons of hazardous air pollutants annually, accounting for roughly 40% of all hazardous air pollutant emissions from point sources. In fact, coal-burning power plants release more hazardous air pollutants than any other point source category. These plants are the largest point source category of hydrochloric acid, mercury, and arsenic releases to air.[1] Additionally, coal-fired plants are major sources of emissions for criteria air pollutants (CAPs) such as sulfur dioxide, oxides of nitrogen, and particulate matter.

1.2 The National Emissions Inventory

The U.S. EPA's National Emission Inventory (NEI) can be found at http://www.epa.gov/ttn/chief/net/1999inventory.html. The EPA's Emission Factor and Inventory Group (EPA/OAR/OAQPS/EMAD/EFIG) prepares a national database of air emissions information with input from numerous state and local air agencies, from tribes, and from industry. This database contains information on stationary and mobile sources that emit criteria air pollutants and their precursors, as well as hazardous air pollutants (HAPs). The database includes estimates of annual emissions, by source, of air pollutants in each area of the country, on an annual basis. The NEI includes emission estimates for all 50 states, the District of Columbia, Puerto Rico, and the Virgin Islands. Emission estimates for individual points or major sources (facilities), as well as county-level estimates for area, mobile, and other sources, are available currently for years 1985 through 1999 for criteria pollutants, and for years 1996 and 1999 for HAPs. Data from the NEI are used for air dispersion modeling purposes, regional strategy development, regulation setting, air toxics risk assessment, and tracking trends in emissions over time. For emission inventories prior to 1999, criteria pollutant emission estimates were maintained in the National Emission Trends (NET) database, and HAP emission estimates were maintained in the National Toxics Inventory (NTI) database. Since 1999, criteria and HAP emissions data are being prepared in a more integrated fashion in the NEI, which takes the place of the NET and the NTI.

1.3 Criteria Air Pollutants

Six criteria air pollutants (CAPs) are tracked on the NEI. They are carbon monoxide, lead, nitrogen dioxide, particulate matter, ozone, and sulfur dioxide.

[1] U.S. EPA, National Emissions Inventory (NEI) 2002: Inventory Data: Point Sector Data—ALLNEI HAP Annual 01232008. 2007. http://www.epa.gov/ttn/chief/net/2002inventory.html#inventorydat

1.3.1 Carbon Monoxide

Carbon monoxide (CO) is a colorless, odorless gas emitted from combustion processes. Nationally, and particularly in urban areas, the majority of CO emissions to ambient air come from mobile sources.

CO can cause harmful health effects by reducing oxygen delivery to the body's organs (such as the heart and brain) and tissues. At extremely high levels, CO can cause death. Exposure to CO can reduce the oxygen-carrying capacity of the blood. People with several types of heart disease already have a reduced capacity for pumping oxygenated blood to the heart, which can cause them to experience myocardial ischemia (reduced oxygen to the heart), often accompanied by chest pain (angina), when exercising or under increased stress. For these people, short-term CO exposure further affects their body's already-compromised ability to respond to the increased oxygen demands of exercise or exertion.

The EPA first set air quality standards for CO in 1971. For protection of both public health and welfare, the EPA set an 8-hour primary standard at 9 parts per million (ppm) and a 1-hour primary standard at 35 ppm. In a review of the standards completed in 1985, the EPA revoked the secondary standards (for public welfare) due to a lack of evidence of adverse effects on public welfare at or near ambient concentrations. The last review of the CO National Ambient Air Quality Standard (NAAQS) was completed in 1994, and the agency chose not to revise the standards at that time.

The Clean Air Act (CAA) requires the EPA to set national ambient air quality standards for "criteria pollutants." Currently, carbon monoxide and five other major pollutants are criteria pollutants; the others are ozone, lead, sulfur oxides, nitrogen oxides, and particulate matter. The law also requires the EPA to review the standards periodically and revise them if appropriate to ensure that they provide the requisite amount of health and environmental protection and to update those standards as necessary.

1.3.2 Lead

Lead (Pb) is a metal found naturally in the environment as well as in manufactured products. The major sources of lead emissions have historically been motor vehicles and industrial sources. As a result of the EPA's regulatory efforts to remove lead from motor vehicle gasoline, emissions of lead from the transportation sector dramatically declined by 95% between 1980 and 1999, and levels of lead in the air decreased by 94% between 1980 and 1999.

Today, the highest levels of lead in air are usually found near lead smelters. The major sources of lead emissions to the air today are ore and metals processing and leaded aviation gasoline. In addition to exposure to lead in air, other major exposure pathways include ingestion of lead in drinking water and lead-contaminated food as well as incidental ingestion of lead-contaminated soil and dust. Lead-based paint remains a major exposure pathway in older homes.

Once taken into the body, lead distributes throughout the circulatory system in the blood and is accumulated in the bones. Depending on the level of exposure, lead can adversely affect the nervous system, kidney function, immune system, reproductive and developmental systems, and the cardiovascular system. Lead exposure also affects the oxygen-carrying capacity of the blood. The lead effects most commonly encountered are neurological effects in children and cardiovascular effects (e.g., high blood pressure and heart disease) in adults. Infants and young children are especially sensitive to even low levels of lead, which may contribute to behavioral problems, learning deficits, and lowered IQ.

Lead is persistent in the environment and accumulates in soils and sediments through deposition from air sources, direct discharge of waste streams to water bodies, mining, and erosion. Ecosystems near point sources of lead demonstrate a wide range of adverse effects including losses in biodiversity, changes in community composition, decreased growth and reproductive rates in plants and animals, and neurological effects in vertebrates.

The Clean Air Act established two types of national air quality standards for lead. These are primary standards and secondary standards. *Primary standards* set limits to protect public health, including the health of "sensitive" populations such as asthmatics, children, and the elderly. *Secondary standards* set limits to protect public welfare, including protection against visibility impairment, damage to animals, crops, vegetation, and buildings. The Clean Air Act requires the EPA to review the latest scientific information and standards every 5 years. Before new standards are established, policy decisions undergo rigorous review by the scientific community, industry, public interest groups, the general public, and the Clean Air Scientific Advisory Committee (CASAC).

1.3.3 Nitrogen Dioxide

Nitrogen dioxide (NO_2) is one of a group of highly reactive gases known as "oxides of nitrogen," or "nitrogen oxides (NOx)." Other nitrogen oxides include nitrous acid and nitric acid. While the EPA's National Ambient Air Quality Standard covers this entire group of NOx, NO_2 is the component of greatest interest and the indicator for the larger group of nitrogen oxides. NO_2

forms quickly from emissions from cars, trucks, and buses; power plants; and off-road equipment. In addition to contributing to the formation of ground-level ozone and fine particle pollution, NO_2 is linked with a number of adverse effects on the respiratory system.

The EPA first set standards for NO_2 in 1971, setting both a primary standard (to protect health) and a secondary standard (to protect the public welfare) at 0.053 parts per million (53 ppb), averaged annually. The agency has reviewed the standards twice since that time but chose not to revise the standards at the conclusion of each review.

Scientific evidence links short-term NO_2 exposures, ranging from 30 minutes to 24 hours, with adverse respiratory effects, including airway inflammation in healthy people and increased respiratory symptoms in people with asthma. Studies show a connection between breathing elevated short-term NO_2 concentrations and increased visits to emergency departments and hospital admissions for respiratory issues, especially asthma.

The sum of nitric oxide (NO) and NO_2 is commonly called nitrogen oxides or NOx. NOx react with ammonia, moisture, and other compounds to form small particles. These small particles penetrate deeply into sensitive parts of the lungs and can cause or worsen respiratory disease, such as emphysema and bronchitis, and can aggravate existing heart disease, leading to increased hospital admissions and premature death. Also, ozone is formed when NOx and volatile organic compounds react in the presence of heat and sunlight. Children, the elderly, people with lung diseases such as asthma, and people who work or exercise outside are at risk for adverse effects from ozone. These effects include reduction in lung function and increased respiratory symptoms as well as respiratory-related emergency department visits, hospital admissions, and possibly premature deaths. Emissions that lead to the formation of NO_2 generally also lead to the formation of other NOx. Emissions control measures leading to reductions in NO_2 can generally be expected to reduce population exposures to all gaseous NOx. This may have the important cobenefit of reducing the formation of ozone and fine particles, both of which pose significant public health threats.

1.3.4 Particulate Matter

"Particulate matter," also known as particle pollution or PM, is a complex mixture of extremely small particles and liquid droplets. Particle pollution is made up of a number of components, including acids (such as nitrates and sulfates), organic chemicals, metals, and soil or dust particles. The size of particles is directly linked to their potential for causing health problems. The EPA is concerned about particles that are 10 micrometers (μm) in diameter or smaller because those are the particles that generally pass through the throat and nose and enter the lungs. Once inhaled, these particles can affect the heart and lungs and cause serious health effects. The EPA groups particle pollution into two categories: "Inhalable coarse particles," such as those found near roadways and dusty industries, are larger than 2.5 μm and smaller than 10 μm in diameter; and "Fine particles," such as those found in smoke and haze, are 2.5 μm in diameter and smaller. These particles can be directly emitted from sources such as forest fires, or they can form when gases emitted from power plants, industries, and automobiles react in the air.

The size of particles is directly linked to their potential for causing health problems. Small particles less than 10 μm in diameter pose the greatest problems because they can get deep into your lungs, and some may even get into your bloodstream. As noted, exposure to such particles can affect both the lungs and heart. The Clean Air Act requires the EPA to set air quality standards to protect both public health and the public welfare (e.g., crops and vegetation). Particle pollution—especially fine particles—contains microscopic solids or liquid droplets that are so small that they can get deep into the lungs and cause serious health problems. Numerous scientific studies have linked particle pollution exposure to a variety of problems, including increased respiratory symptoms, such as irritation of the airways, coughing, or difficulty breathing decreased lung function; aggravated asthma; development of chronic bronchitis; irregular heartbeat; nonfatal heart attacks; and premature death in people with heart or lung disease. People with heart or lung diseases, children, and older adults are the most likely to be affected by particle pollution exposure. The EPA warns that even if a person is healthy, he or she may experience temporary symptoms from exposure to elevated levels of particle pollution.

The nation's air quality standards for particulate matter were first established in 1971 and were not significantly revised until 1987, when the EPA changed the indicator of the standards to regulate inhalable particles smaller than, or equal to, 10 μm in diameter (that's about one-fourth the size of a single grain of table salt). Ten years later, after a lengthy review, the EPA revised the PM standards, setting separate standards for fine particles ($PM_{2.5}$) based on their link to serious health problems ranging from increased symptoms, hospital admissions, and emergency room visits for people with heart and lung disease, to premature death in people with heart or lung disease. The 1997 standards also retained but slightly revised standards for

PM_{10}, which were intended to regulate "inhalable coarse particles" that ranged from 2.5 to 10 μm in diameter. PM_{10} measurements, however, contain both fine and coarse particles. The EPA revised the air quality standards for particle pollution in 2006. The 2006 standards tighten the 24-hour fine particle standard from the current level of 65 micrograms per cubic meter ($\mu g/m^3$) to 35 $\mu g/m^3$ and retain the current annual fine particle standard at 15 $\mu g/m^3$. The EPA decided to retain the existing 24-hour PM_{10} standard of 150 $\mu g/m^3$. The EPA revoked the annual PM_{10} standard because available evidence does not suggest a link between long-term exposure to PM_{10} and health problems.

1.3.5 Ozone

Ozone (O_3) is a gas composed of three oxygen atoms. It is not usually emitted directly into the air, but at ground level is created by a chemical reaction between oxides of nitrogen (NOx) and volatile organic compounds (VOCs) in the presence of sunlight. Ozone has the same chemical structure whether it occurs miles above the earth or at ground level and can be "good" or "bad," depending on its location in the atmosphere. In the earth's lower atmosphere, ground-level ozone is considered "bad." Motor vehicle exhaust and industrial emissions, gasoline vapors, and chemical solvents as well as natural sources emit NOx and VOCs that aid in the formation of ozone.

Ground-level ozone is the primary constituent of smog. Sunlight and hot weather cause ground-level ozone to form in harmful concentrations. This is known as a summertime air pollutant. Many urban areas tend to have high levels of "bad" ozone, but even rural areas are also subject to increased ozone levels because wind carries ozone and pollutants that form bad ozone hundreds of miles away from their original sources. The so-called "good" ozone occurs naturally in the stratosphere approximately 10 to 30 miles above the earth's surface and forms a layer that protects life on earth from the sun's harmful rays.

Breathing ozone can trigger a variety of health problems including chest pain, coughing, throat irritation, and congestion. It can worsen bronchitis, emphysema, and asthma. Ground-level ozone also can reduce lung function and inflame the linings of the lungs. Repeated exposure may permanently scar lung tissue. The Clean Air Act requires the EPA to set air quality standards to protect both public health and the public welfare (e.g., crops and vegetation). Ground-level ozone affects both. People with lung disease, children, older adults, and people who are active can be affected when ozone levels are unhealthy. Numerous scientific studies have linked ground-level ozone exposure to a variety of problems, including airway irritation, coughing, and pain when taking a deep breath; wheezing and breathing difficulties during exercise or outdoor activities; inflammation, which is much like a sunburn on the skin; aggravation of asthma and increased susceptibility to respiratory illnesses such as pneumonia and bronchitis; and, permanent lung damage with repeated exposures. There are both primary and secondary standards (described in the next section).

1.3.6 Sulfur Dioxide

Sulfur dioxide (SO_2) is one of a group of highly reactive gases known as "oxides of sulfur." The largest sources of SO_2 emissions are from fossil fuel combustion at power plants (73%) and other industrial facilities (20%). Smaller sources of SO_2 emissions include industrial processes such as extracting metal from ore, and the burning of high-sulfur-containing fuels by locomotives, large ships, and nonroad equipment. SO_2 is linked with a number of adverse effects on the respiratory system.

The EPA first set standards for SO_2 in 1971. It set a 24-hour primary standard at 140 ppb and an annual average standard at 30 ppb (to protect health). The EPA also set a 3-hour average secondary standard at 500 ppb (to protect the public welfare). The last review of the SO_2 NAAQS was completed in 1996, and the EPA chose not to revise the standards. In the last review, the EPA also considered, but did not set, a 5-minute NAAQS to protect asthmatics at elevated ventilation rates from bronchoconstriction and respiratory symptoms associated with 5–10-minute peaks of SO_2. Scientific evidence links short-term exposures to SO_2, ranging from 5 minutes to 24 hours, with an array of adverse respiratory effects including bronchoconstriction and increased asthma symptoms. These effects are particularly important for asthmatics at elevated ventilation rates (e.g., while exercising or playing). Studies also show a connection between short-term exposure and increased visits to emergency departments and hospital admissions for respiratory illnesses, particularly in at-risk populations including children, the elderly, and asthmatics. The EPA's NAAQS for SO_2 is designed to protect against exposure to the entire group of sulfur oxides (SOx). SO_2 is the component of greatest concern and is used as the indicator for the larger group of gaseous sulfur oxides (SOx). Other gaseous sulfur oxides (e.g., SO_3) are found in the atmosphere at concentrations much lower than those of SO_2. Emissions that lead to high concentrations of SO_2 generally also lead to the formation of other forms of SOx. Control measures that reduce SO_2 can generally be expected to reduce people's exposures to all gaseous SOx. This may have the important cobenefit of reducing the formation of fine sulfate particles, which pose significant public health threats. SOx can react with other compounds in the atmosphere to form small particles. These particles penetrate deeply into sensitive parts of the lungs and can cause or worsen respiratory disease, such as emphysema and bronchitis, and can aggravate existing heart disease, leading to increased hospital admissions and premature death. The EPA's NAAQS for particulate matter are designed to provide protection against these health effects.

1.3.7 NAAQS

Table 1.1 reports the National Ambient Air Quality Standards for the six criteria air pollutants.

The units of measure for the standards reported in Table 1.1 are parts per million (ppm) by volume, parts per billion (ppb—1 part in 1,000,000,000) by volume, milligrams per cubic meter of air (mg/m^3), and micrograms per cubic meter of air ($\mu g/m^3$).

1.4 Comparing Emissions

1.4.1 National Emissions Inventory

Table 1.2 provides a tabulation of the 2005 reported NEI released from 95 coal-fired plants across the country. The magnitude of the emissions is indeed impressive, particularly in light of the fact that there are more than 300 such plants across the United States. Some of the facilities do not report emissions (ND = No Data). In terms of total releases, the plants identified in Table 1.2 released more than 7.2 billion pounds of criteria air pollutants, composed of nearly 6.7 billion pounds of SOx, more than 337 million pounds of PM_{10}, more than 252 million pounds of CO, 21.5 million pounds of VOCs, and 8.3 million pounds of ammonia (NH_3). The releases reported in Table 1.2 are criteria air pollutants.

1.4.2 Toxics Release Inventory

In 1984, a deadly cloud of methyl isocyanate killed thousands of people in Bhopal, India. Shortly thereafter, there was a serious chemical release at a sister plant in West Virginia. These incidents underscored demands by industrial workers and communities in several states for information on hazardous materials. Public interest and environmental organizations accelerated demands for information on toxic chemicals being released "beyond

Table 1.1 National Ambient Air Quality Standards

	Primary Standards		Secondary Standards	
Pollutant	**Level**	**Averaging Time**	**Level**	**Averaging Time**
Carbon Monoxide	9 ppm (10 mg/m^3)	8-hour [(1)]	None	
	35 ppm (40 mg/m^3)	1-hour [(1)]		
Lead	0.15 μg/m^3 [(2)]	Rolling 3-Month Average	Same as Primary	
Nitrogen Dioxide	53 ppb [(3)]	Annual (Arithmetic Average)	Same as Primary	
	100 ppb	1-hour [(4)]	None	
Particulate Matter (PM_{10})	150 μg/m^3	24-hour [(5)]	Same as Primary	
Particulate Matter ($PM_{2.5}$)	15.0 μg/m^3	Annual [(6)] (Arithmetic Average)	Same as Primary	
Ozone	0.075 ppm (2008 std)	8-hour [(7)]	Same as Primary	
	0.08 ppm (1997 std)	8-hour [(8)]	Same as Primary	
	0.12 ppm	1-hour [(9)]	Same as Primary	
Sulfur Dioxide	0.03 ppm	Annual (Arithmetic Average)	0.5 ppm	3-hour [(1)]
	0.14 ppm	24-hour [(1)]		
	75 ppb [(10)]	1-hour	None	

[(1)]Not to be exceeded more than once per year.
[(2)]Final rule signed October 15, 2008. The 1978 lead standard (1.5 μg/m^3 as a quarterly average) remains in effect until 1 year after an area is designated for the 2008 standard, except that in areas designated nonattainment for the 1978 standard, the 1978 standard remains in effect until implementation plans to attain or maintain the 2008 standard are approved.
[(3)]The official level of the annual NO_2 standard is 0.053 ppm, equal to 53 ppb, which is shown here for the purpose of clearer comparison to the 1-hour standard.
[(4)]To attain this standard, the 3-year average of the 98th percentile of the daily maximum 1-hour average at each monitor within an area must not exceed 100 ppb (effective January 22, 2010).
[(5)]Not to be exceeded more than once per year on average over 3 years.
[(6)]To attain this standard, the 3-year average of the weighted annual mean $PM_{2.5}$ concentrations from single or multiple community-oriented monitors must not exceed 15.0 μg/m^3.
[(7)]To attain this standard, the 3-year average of the fourth-highest daily maximum 8-hour average ozone concentrations measured at each monitor within an area over each year must not exceed 0.075 ppm (effective May 27, 2008).
[(8)](a) To attain this standard, the 3-year average of the fourth-highest daily maximum 8-hour average ozone concentrations measured at each monitor within an area over each year must not exceed 0.08 ppm.
(b) The 1997 standard—and the implementation rules for that standard—will remain in place for implementation purposes as the EPA undertakes rulemaking to address the transition from the 1997 ozone standard to the 2008 ozone standard.
(c) The EPA is in the process of reconsidering these standards (set in March 2008).
[(9)](a) The EPA revoked the 1-hour ozone standard in all areas, although some areas have continuing obligations under that standard ("anti-backsliding").
(b) The standard is attained when the expected number of days per calendar year with maximum hourly average concentrations above 0.12 ppm is < 1.
[(10)](a) Final rule signed June 2, 2010. To attain this standard, the 3-year average of the 99th percentile of the daily maximum 1-hour average at each monitor within an area must not exceed 75 ppb.
Source: http://www.epa.gov/air/criteria.html

Table 1.2 2005 NEI Reported Emissions in Tons

Facility	State	VOC	NH_3	SOx	CO	Primary PM_{10}	Total
Gibson Generating Station—Duke Energy Owensville (Also referred to as PSI Energy—Gibson)	IN	315	654	154,259	2,625	6,016	163,870
Gaston Steam Plant	AL	173	71	127,705	1,483	7,179	136,611
Cardinal Plant	OH	137	3	115,848	1,142	4,837	121,966
Amos Plant—American Electric Power (Also referred to as Appalachian Power)	WV	221	4	112,415	1,841	4,538	119,018
Detroit Edison Monroe Power—DTE Energy	MI	259	4	110,205	2,153	3,356	115,976
Conesville Power Plant	OH	130	2	106,586	1,075	6,709	114,502
Wansley Steam Plant—Georgia Power	GA	188	129	101,547	1,829	5,035	108,728
Roxboro Steam Plant—Progress Energy	NC	173	2	103,186	1,443	3,108	107,912
Kammer and Mitchel Plants (includes two plants)	WV	168	2	96,401	1,082	224	97,877
Branch Steam Electric Generating Plant	GA	133	57	90,505	952	5,790	97,437
Scherer Steam—Georgia Power	GA	417	179	82,895	2,981	3,529	90,001
Paradise Fossil Plant	KY	215	92	84,399	1,842	2,438	88,987
Gorgas Steam Plant	AL	119	46	84,053	1,018	3,344	88,580
Chesterfield Power Station—Dominion	VA	102	2	78,501	824	4,620	84,048
Kyger Creek Station	OH	108	1	72,543	772	5,063	78,487
Wabash River Generating Station—Duke Energy	IN	462	227	66,813	6,484	3,622	77,607
Jeffrey Energy Center—Westar Energy	KS	293	147	69,565	2,432	3,307	75,745
Yates Steam Generating Plant	GA	105	50	66,533	733	3,499	70,920

Table 1.2 2005 NEI Reported Emissions in Tons—*cont'd*

Facility	State	VOC	NH_3	SOx	CO	Primary PM_{10}	Total
Rockport Plant—American Electric Power	IN	287	4	67,187	2,399	1,000	70,876
Kingston Fossil Plant	TN	115	63	56,222	960	3,425	60,784
Labadie Power Station	MO	298	152	55,502	2,484	2,065	60,501
Gallagher Generating Station—Duke Energy	IN	50	1	56,667	358	2,586	59,662
Barry Steam Plant—Alabama Power	AL	212	145	53,657	1,580	3,004	58,598
Miller Steam Plant—Alabama Power	AL	440	186	50,802	3,756	2,042	57,226
Ghent Generating Station—E. ON US	KY	197	85	50,149	1,690	2,461	54,582
Sioux Power Station—Ameren UE	MO	185	90	51,265	865	1,498	53,903
Big Sandy Plant—American Electric Power	KY	100	45	50,071	753	2,713	53,683
G.G. Allen Steam Plant	NC	78	2	45,425	2,258	2,264	50,027
Tanners Creek Plant—American Electric Power	IN	88	1	46,536	534	2,241	49,399
Green County Steam Plant	AL	73	30	45,607	614	2,632	48,956
Big Cajun II—NRG Energy (Also referred to as LA Generating LLC/BI)	LA	218	110	42,329	1,815	4,253	48,725
HW Pirkey Power Plant—American Electric Power	TX	65	64	11,699	34,005	1,421	47,254
Hammond Steam Generating Station	GA	62	26	39,549	444	1,864	41,945
Petersburg Generating Station	IN	179	2	37,654	1,281	1,943	41,059
Shawnee Fossil Plant	KY	146	59	37,739	1,626	783	40,353
Widows Creek Fossil Plant—Tennessee Valley Authority	AL	123	62	34,382	1,027	3,232	38,825
Brayton Point Station	MA	100	22	32,519	1,801	1,752	36,195
Coal Creek Station	ND	154	ND	27,390	1,930	2,945	32,419

(*Continued*)

Table 1.2 2005 NEI Reported Emissions in Tons—*cont'd*

Facility	State	VOC	NH_3	SOx	CO	Primary PM_{10}	Total
Milton R. Young Station—Minnkota Power Cooperative	ND	227	66	29,407	1,033	1,304	32,037
Karn and Weadock Generating Plants	MI	108	15	29,944	942	811	31,819
Rush Island Power Station—Ameren UE	MO	159	81	28,385	1,327	1,065	31,017
Cliffside Steam Station—Duke Energy	NC	45	0	28,221	377	2,196	30,840
La Cygne Generating Station	KS	214	80	27,630	1,326	1,024	30,274
McDonough Atkinson Steam Plant—Georgia Power	GA	50	24	27,606	352	1,720	29,753
Baldwin Energy Station—Dynegy	IL	332	2	24,009	1,817	2,751	28,911
George Neal Station (North)—MidAmerican Energy	IA	143	63	24,244	1,034	2,800	28,284
Sherburne Co. Xcel Energy	MN	248	8	22,838	2,071	2,479	27,644
Gallatin Fossil Plant	TN	112	52	25,907	923	647	27,641
Cholla Power Plant—Arizona Public Service Co.	AZ	6	59	22,028	1,743	2,989	26,826
Naughton Power Plant	WY	96	42	23,229	827	989	25,183
Allen Fossil Plant	TN	137	5	23,889	622	481	25,134
Boswell Energy Center—Minnesota Power	MN	64	1	19,650	1,059	2,644	23,417
L.V. Sutton Electric Plant—Progress Energy	NC	41	21	21,146	340	844	22,391
Killen Generating Station	OH	44	1	19,566	366	735	20,711
Iatan Generating Station—Great Plains Energy	MO	75	39	19,217	627	603	20,561
Canadys Station—SCE&G SCANA	SC	30	14	19,558	251	260	20,114
Lansing Smith Generating Plant	FL	34	62	18,516	262	1,049	19,923

Table 1.2 2005 NEI Reported Emissions in Tons—*cont'd*

Facility	**State**	**VOC**	**NH_3**	**SOx**	**CO**	**Primary PM_{10}**	**Total**
Kincaid Generating Station	IL	204	0	17,461	926	553	19,144
Laramie River Station	WY	257	129	13,098	2,145	2,492	18,121
Montrose Generating Station	MO	60	30	15,709	502	525	16,826
Asheville Plant—Progress Energy	NC	31	2	15,545	240	416	16,234
Frank E. Rats Generating Station—Hoosier Energy	IN	14	0	15,123	119	801	16,058
Lee Steam Plant—Progress Energy	NC	33	35	15,020	250	464	15,803
New Madrid Power Plant	MO	219	59	13,702	991	727	15,698
Coleto Creek Power Station	TX	73	37	14,396	601	463	15,569
Cape Fear Steam Plant—Progress Energy	NC	27	0	13,633	219	1,053	14,933
Meredosia Power Station—Ameren	IL	25	2	11,980	179	1,110	13,296
Louisa Generating Station	IA	0	34	12,326	5	370	12,735
Asbury Generating Station	MO	26	11	11,965	184	253	12,439
W.S. Lee Steam Station—Duke Energy	SC	22	0	11,228	195	875	12,320
Hugo Plant—Western Farmers Electric	OK	59	30	10,323	1,478	105	11,995
Stanton Energy Center—Orlando Utilities Commission	FL	105	43	8,838	915	921	10,822
Buck Steam Station—Duke Energy	NC	23	0	9,583	185	504	10,296
Gadsden Steam Plant—Alabama Power	AL	9	4	9,343	78	494	9,930
Flint Creek Power—AEP	AR	64	32	8,227	529	297	9,149
Havana Power Station—Ameren	IL	41	21	7,152	342	641	8,197

(*Continued*)

Table 1.2 2005 NEI Reported Emissions in Tons—*cont'd*

Facility	State	VOC	NH_3	SOx	CO	Primary PM_{10}	Total
Intermountain Power Station	UT	14	83	3,597	1,384	1,920	6,998
Twin Oaks Power Station	TX	2	33	5,482	902	477	6,896
Trimble County Generating Station	KY	44	23	5,235	369	461	6,133
Duck Creek Station	IL	23	0	5,134	189	341	5,687
Riverton Power Plant—Empire District Electric Co. (Also referred to as Petersburg Generation)	KS	10	0	4,357	83	166	4,616
E.W. Brown Generating Station	KY	41	21	3,666	339	439	4,506
Apache Generating Station—Arizona Electric Power Cooperative	AZ	51	25	2,657	468	609	3,810
Pearl Station	IL	3	0	1,698	29	2	1,732
Dolet Hills Power Station—Cleco	LA	17	ND	15	43	1,235	1,310
Mayo Generating Plant	NC	75	ND	10	103	51	240
Welsh Power Plant—American Electric Power	TX	16	ND	72	32	3	124
Okluanion Power Station	TX	4	ND	4	21	53	82
BP Amoco Council Bluffs Terminal (Council Bluffs Energy Center)	IA	72	ND	ND	3	6	81
Muskingum River Plant	OH	1	18	1	44	2	66
J.M. Stuart Station—Dayton Power & Light	OH	2	ND	ND	ND	ND	2
Clifty Creek Station	IN	ND	ND	ND	ND	ND	-
Leland Olds Station—Basin Electric Power	ND	ND	ND	ND	ND	ND	-
Meramec Power Plant	MO	ND	ND	ND	ND	ND	-
Wood River Station—Dynegy	IL	ND	ND	ND	ND	ND	-
Total		**10,724**	**4,131**	**3,337,880**	**126,280**	**168,560**	**3,647,575**

the fence line"—outside the facility. Against this background, the Emergency Planning and Community Right-to-Know Act (EPCRA) was enacted in 1986. EPCRA's primary purpose is to inform citizens of toxic chemical releases in their areas. EPCRA Section 313 requires the EPA and the states to collect data annually on releases and transfers of certain toxic chemicals from industrial facilities and make the data available to the public through the Toxics Release Inventory (TRI).

Until 1998 power plants were exempt from reporting their releases to the TRI. Now more than a decade's worth of reporting data has been made available for communities and researchers to examine.

A problem with the TRI is that the reporting requirements are limited to specific chemical or chemical classes. There are about 650 chemicals on which facilities are required to report their releases; however, they are only required to report estimated releases and transfers that exceed a threshold amount for that chemical. Generally, this is 10,000 pounds of a chemical within a reporting year or 25,000 pounds of an aggregate class. A second and perhaps more serious problem with TRI reporting is that the reported values are entirely calculated. There are no measured releases to air or water, although generally off-site transfers of wastes are based on actual bills of lading. There is considerable latitude in developing emissions inventories for the purposes of TRI reporting, and as such TRI reported emissions.

An example of the discrepancies found in reporting is taken from an examination of the Tennessee Valley Authority (TVA) Kingston Tennessee plant. Table 1.3 tabulates the reported air emissions for the facility obtained from the EPA's ToxNet website. According to the TRI, TVA reports fugitive emissions to be almost insignificant (on the average about 0.12%). But according to the 2005 NEI database, about 5% of the total emissions from the facility are fugitive emissions, and all of these are in the form of PM, which facilities are not required to report to the TRI. Despite concerns over the reported releases, the TRI does represent a comprehensive database of reported releases to multimedia.

Table 1.4 reports the releases to air for 95 power plants. This table reports the releases to air in pounds for TRI-listed chemicals. The table shows a trend in air emissions reduction between 1998 and 2009. The last column of the table reports the percent reduction in air emissions over the base year (first year of the facility's reporting to the TRI). This trend can be explained by two distinct yet complementary interpretations. First, the industry sector from this small sampling shows dramatic reductions in HAP emissions, which should be a reflection of better, more improved pollution controls and investments into pollution prevention practices and

Table 1.3 TRI Reported Releases to Air

Chemical Name	Unit	2009	2008	2007	2006	2005	2004	2003	2002	2001	2000	1999	1998	Total
Fugitive Releases to Air														
1,2,4-Trimethylbenzene	Pounds	-	-	-	-	-	-	-	250	250	250	250	250	1,250
Ammonia	Pounds	2,600	11,000	7,000	12,000	11,000	1,300	-	5	-	-	-	-	44,905
Arsenic	Pounds	-	5	5	5	5	5	5	5	5	5	5	5	55
Barium	Pounds	7	250	250	250	250	250	250	250	250	250	250	250	2,757
Benzo(G,H,I) Perylene	Pounds	-	-	-	-	0	-	-	-	-	-	-	-	0
Chromium	Pounds	-	250	5	5	5	250	250	250	5	5	5	5	1,035
Cobalt	Pounds	-	5	5	5	5	5	5	5	5	5	5	5	55
Copper	Pounds	1	5	5	5	5	5	5	5	5	5	5	5	56
Hydrochloric Acid	Pounds	-	-	-	-	5	5	5	5	5	5	5	5	40
Hydrogen Fluoride	Pounds	-	-	-	-	5	5	5	5	5	5	5	5	40
Lead	Pounds	3	5	2	2	3	6	4	4	2	5	5	5	46
Manganese	Pounds	71	250	250	250	250	250	250	250	250	250	250	250	2,821
Mercury	Pounds	-	-	-	-	0	0	0	0	-	0	-	-	1
N-Hexane	Pounds	-	-	-	-	-	-	-	250	250	250	250	250	1,250
Nickel	Pounds	-	250	5	5	5	250	5	5	5	5	5	5	545
PACs*	Pounds	0.1	0.1	0.1	0.1	0.77	0.8	0.8	0.8	0.4	0.8	0	0	4.77
Selenium	Pounds	-	-	-	-	-	5	5	5	5	5	5	5	35
Styrene	Pounds	-	5,400	-	-	-	-	-	-	-	-	-	-	5,400
Sulfuric Acid	Pounds	-	-	-	-	5	5	5	5	5	5	5	250	285
Vanadium	Pounds	2	5	5	5	5	5	5	5	5	5	-	-	47
Zinc	Pounds	1	5	5	5	5	5	5	5	5	5	5	5	56
Total Air Fugitives	Pounds	2,685.1	17,429.6	7,537.3	12,537	11,554	2,352	805	1,309.8	1,057.8	1,060.9	1,055.0	1,300.0	60,683.7
Stack Releases To Air														
1,2,4-Trimethylbenzene	Pounds	0	0	0	0	0	0	0	250	250	250	250	250	1250
Ammonia	Pounds	4,800	27,000	14,000	23,000	18,000	-	-	-	-	-	-	-	86,800
Arsenic	Pounds	-	250	250	250	250	250	250	250	1,500	2,000	2,100	2,100	9,450
Barium	Pounds	460	2,900	2,400	1,600	1,500	750	750	750	1,000	1,800	1,800	1,900	17,610

Benzo(G,H,I) Perylene	Pounds	-	0	0	0	0	0	0	0	0	0	-	-	1
Chromium	Pounds	-	250	250	250	250	250	750	750	750	250	750	750	5,250
Cobalt	Pounds	-	250	250	250	250	250	250	250	250	250	250	250	2,750
Copper	Pounds	89	250	250	250	250	250	250	250	750	750	750	750	4,839
Hydrochloric Acid	Tons	385	1,550	1,700	1,850	1,650	2,100	2,250	2,300	2,000	2,100	2,150	3,050	23,085
Hydrogen Fluoride	Tons	48	225	225	250	225	245	250	265	255	270	275	300	2,833
Lead	Pounds	61	280	290	240	220	310	330	380	240	250	250	250	3,101
Manganese	Pounds	140	750	750	750	750	750	750	750	750	250	250	250	6,890
Mercury	Pounds	100	480	480	480	430	450	490	480	450	510	-	-	4,350
N-Hexane	Pounds	-	-	-	-	-	-	-	250	250	250	250	250	1,250
Naphthalene	Pounds	13	250	250	250	250	250	-	-	-	-	-	-	1,263
Nickel	Pounds	-	250	250	250	250	250	750	750	250	250	250	250	3,750
PACs*	Pounds	7	32	32	32	29	27	28	30	27	29	-	-	273
Selenium	Pounds	-	-	9,000	10,000	-	11,000	12,000	12,000	7,700	9,100	9,200	9,400	89,400
Styrene	Pounds	-	250	-	-	-	-	-	-	-	-	-	-	250
Sulfuric Acid	Tons	100	285	330	340	390	500	425	1,100	700	420	500	550	5,640
Vanadium	Pounds	92	250	250	250	250	250	750	750	750	750	-	-	4,342
Zinc	Pounds	320	1,500	1,500	1,400	1,300	1,500	1,600	1,700	250	250	250	250	11,820
Total Air Stack (Excluding Dioxins)	Tons	536.0	2,077.5	2,270.1	2,459.6	2,277.0	2,853.3	2,934.5	3,674.8	2,962.6	2,798.5	2,933.2	3,908.3	31,685.3
Total Air (Stack + Fugitive; Excluding Dioxins)	Tons	537.4	2,086.2	2,273.9	2,465.9	2,282.8	2,854.4	2,934.9	3,675.4	2,963.1	2,799.0	2,933.7	3,909.0	31,715.7
Percent of Releases As Fugitive	Percent	0.25	0.42	0.17	0.25	0.25	0.04	0.01	0.02	0.02	0.02	0.02	0.02	0.10
Dioxin – Fugitive Releases	Grams	NR	NR	NR	NR	0.18	0.18	0.18	0.18	0.08	0.08	NR	NR	0.98
Dioxin – Stack Releases	Grams	0.45	2.24	2.20	2.20	2.00	1.9	2.0	2.05	1.9	3.9	NR	NR	20.8

*PACs—Polycyclic Aromatic Compounds; NR – Not Reported
Source: ToxNet

Table 1.4 TRI Total Air Releases in Pounds

Facility	State	Total Air (All Chemicals) (Pounds)	Base Year Amt. (Pounds)	Base Year	End Year	% Reduction over Base Year
Amos Plant—American Electric Power	WV	186,493,849	16,709,646	1998	2009	91.04
Roxboro Steam Plant—Progress Energy	NC	149,551,536	15,148,574	1998	2009	89.87
Kammer and Mitchel Plants	WV	136,910,758	16,801,439	1998	2009	87.73
Detroit Edison Monroe Power—DTE Energy	MI	122,860,698	8,058,508	1998	2009	93.44
J.M. Stuart Station—Dayton Power & Light	OH	122,543,295	12,807,337	1998	2009	89.55
Cardinal Plant	OH	104,276,033	11,323,277	1998	2009	89.14
Branch Steam Electric Generating Plant	GA	99,782,588	5,576,860	1998	2009	94.41
Muskingum River Plant	OH	95,802,518	7,930,873	1998	2009	91.72
Wansley Steam Plant—Georgia Power	GA	90,215,234	6,675,128	1998	2009	92.60
Paradise Fossil Plant	KY	84,982,393	9,095,350	1998	2009	89.30
Gibson Generating Station—Duke Energy Owensville	IN	81,534,331	8,066,947	1998	2009	90.11
Ghent Generating Station—E. ON US	KY	77,522,146	8,471,776	1998	2009	89.07
Big Sandy Plant—American Electric Power	KY	75,603,735	6,464,315	1998	2009	91.45
Scherer Steam—Georgia Power	GA	68,760,787	8,917,138	1998	2009	87.03
Kyger Creek Station	OH	68,254,048	7,707,645	1998	2009	88.71
Conesville Power Plant	OH	68,176,975	6,132,623	1998	2009	91.00

G.G. Allen Steam Plant	NC	66,393,847	4,347,800	1998	2009	93.45
Kingston Fossil Plant	TN	63,431,322	7,817,950	1998	2009	87.67
Mayo Generating Plant	NC	59,542,212	4,720,134	1998	2009	92.07
Gaston Steam Plant	AL	55,518,001	4,065,426	1998	2009	92.68
Chesterfield Power Station—Dominion	VA	51,197,618	3,936,278	1998	2009	92.31
Green County Steam Plant	AL	48,320,511	4,760,319	1998	2009	90.15
Cliffside Steam Station—Duke Energy	NC	46,483,449	3,920,530	1998	2009	91.57
Tanners Creek Plant—American Electric Power	IN	45,993,636	5,043,274	1998	2009	89.03
Clifty Creek Station	IN	43,452,510	4,199,470	1998	2009	90.34
Barry Steam Plant—Alabama Power	AL	37,451,448	5,637,540	1998	2009	84.95
Lansing Smith Generating Plant	FL	37,393,203	3,279,421	1998	2009	91.23
Gorgas Steam Plant	AL	37,097,776	3,685,888	1998	2009	90.06
Rockport Plant—American Electric Power	IN	36,901,464	2,228,121	1998	2009	93.96
L.V. Sutton Electric Plant—Progress Energy	NC	36,890,064	3,127,286	1998	2009	91.52
E.W. Brown Generating Station	KY	35,561,907	3,704,750	1998	2009	89.58
Widows Creek Fossil Plant—Tennessee Valley Authority	AL	35,262,078	2,728,300	1998	2009	92.26
Gallagher Generating Station—Duke Energy	IN	34,094,275	2,901,835	1998	2009	91.49
Yates Steam Generating Plant	GA	32,714,671	2,529,202	1998	2009	92.27
Karn and Weadock Generating Plants	MI	32,160,334	3,545,292	1998	2009	88.98
Killen Generating Station	OH	30,949,599	4,461,970	1998	2009	85.58
Petersburg Generating Station	IN	30,818,287	2,784,783	1998	2009	90.96
Sioux Power Station—Ameren UE	MO	29,011,986	3,349,974	1998	2009	88.45

(*Continued*)

Table 1.4 TRI Total Air Releases in Pounds—*cont'd*

Facility	State	Total Air (All Chemicals) (Pounds)	Base Year Amt. (Pounds)	Base Year	End Year	% Reduction over Base Year
Baldwin Energy Station—Dynegy	IL	28,065,347	8,450,555	1998	2009	69.89
McDonough Atkinson Steam Plant—Georgia Power	GA	26,975,981	3,021,517	1998	2009	88.80
Hammond Steam Generating Station	GA	26,245,933	1,262,847	1998	2009	95.19
Lee Steam Plant—Progress Energy	NC	23,333,259	1,870,976	1998	2009	91.98
Cape Fear Steam Plant—Progress Energy	NC	22,892,240	2,067,218	1998	2009	90.97
Wabash River Generating Station—Duke Energy	IN	21,652,525	1,767,880	1998	2009	91.84
Gallatin Fossil Plant	TN	20,244,779	3,230,180	1998	2009	84.04
Buck Steam Station—Duke Energy	NC	20,174,814	1,642,715	1998	2009	91.86
Canadys Station—SCE&G SCANA	SC	19,825,447	1,163,099	1998	2009	94.13
Asheville Plant- Progress Energy	NC	18,178,010	2,419,423	1998	2009	86.69
Frank E. Rats Generating Station—Hoosier Energy	IN	18,079,863	1,113,840	1998	2009	93.84
Miller Steam Plant—Alabama Power	AL	17,033,565	2,603,215	1998	2009	84.72
Meramec Power Plant	MO	15,994,354	3,535,476	1998	2009	77.90
W.S. Lee Steam Station—Duke Energy	SC	15,409,091	1,088,320	1998	2009	92.94
Brayton Point Station	MA	14,583,691	1,176,137	1998	2009	91.94
Allen Fossil Plant	TN	14,186,811	410,530	1998	2009	97.11
Big Cajun II—NRG Energy	LA	10,936,032	3,336,934	1998	2009	69.49
Labadie Power Station	MO	10,782,771	1,966,117	1998	2009	81.77
Trimble County Generating Station	KY	10,608,440	901,249	1998	2009	91.50

Shawnee Fossil Plant	KY	10,421,854	—	1994	2009	100.00
Meredosia Power Station—Ameren	IL	9,418,765	970,911	1998	2009	89.69
Wood River Station—Dynegy	IL	9,240,958	967,000	1998	2009	89.54
George Neal Station (North)—MidAmerican Energy	IA	8,642,141	718,535	1998	2009	91.69
Kincaid Generating Station	IL	7,459,843	2,617,278	1998	2009	64.92
Stanton Energy Center—Orlando Utilities Commission	FL	6,951,186	251,710	1998	2009	96.38
Cholla Power Plant—Arizona Public Service Co.	AZ	6,040,663	578,935	1998	2009	90.42
Louisa Generating Station	IA	5,851,325	1,386,290	1998	2009	76.31
Welsh Power Plant—American Electric Power	TX	4,757,921	1,167,030	1998	2009	75.47
Coal Creek Station	ND	4,566,983	665,075	1998	2009	85.44
Asbury Generating Station	MO	3,952,616	251,987	1998	2009	93.62
Rush Island Power Station—Ameren UE	MO	3,921,395	349,294	1998	2009	91.09
New Madrid Power Plant	MO	3,841,275	332,130	1998	2009	91.35
La Cygne Generating Station	KS	3,660,621	284,305	1998	2009	92.23
Dolet Hills Power Station—Cleco	LA	3,651,316	569,959	1998	2009	84.39
Jeffrey Energy Center—Westar Energy	KS	3,614,082	402,165	1998	2009	88.87
Gadsden Steam Plant—Alabama Power	AL	2,989,458	176,971	1998	2009	94.08
Havana Power Station—Ameren	IL	2,909,194	332,965	1998	2009	88.55
Hugo Plant—Western Farmers Electric	OK	2,859,557	270,925	1998	2009	90.53
Coleto Creek Power Station	TX	2,669,319	765,030	1998	2009	71.34

(*Continued*)

Table 1.4 TRI Total Air Releases in Pounds—*cont'd*

Facility	State	Total Air (All Chemicals) (Pounds)	Base Year Amt. (Pounds)	Base Year	End Year	% Reduction over Base Year
Iatan Generating Station—Great Plains Energy	MO	2,639,945	248,065	1998	2009	90.60
HW Pirkey Power Plant—American Electric Power	TX	2,614,525	122,867	1998	2009	95.30
Laramie River Station	WY	2,423,031	1,428,098	1998	2009	41.06
Milton R. Young Station—Minnkota Power Cooperative	ND	2,339,972	182,289	1998	2009	92.21
Riverton Power Plant—Empire District Electric Co.	KS	2,057,020	122,018	1998	2009	94.07
Montrose Generating Station	MO	2,053,530	176,470	1998	2009	91.41
Naughton Power Plant	WY	1,883,703	238,964	1998	2009	87.31
Intermountain Power Station	UT	1,830,291	150,300	1998	2009	91.79
Boswell Energy Center—Minnesota Power	MN	1,657,759	124,200	1998	2009	92.51
Leland Olds Station—Basin Electric Power	ND	1,613,947	142,701	1998	2009	91.16
Flint Creek Power—AEP	AZ	1,417,778	434,817	1998	2009	69.33
Sherburne Co. Xcel Energy	MN	1,310,820	121,446	1998	2009	90.74
Apache Generating Station—Arizona Electric Power Cooperative	AZ	1,087,371	70,750	1987	2009	93.49
Okluanion Power Station	TX	1,066,841	613,546	1998	2009	42.49
Twin Oaks Power Station	TX	824,099	66,893	1998	2009	91.88
Pearl Station	IL	530,612	142,248	1998	2009	73.19
BP Amoco Council Bluffs Terminal (Council Bluffs Energy Center)	IA	5,678	1,831	1998	2000	67.75

source reduction. The data also suggest that, historically, power plants have caused significant levels of air pollution that have impacted community health.

Table 1.5 provides some additional comparisons for the 95 plants considered in this survey. This table gives a breakdown of the reported releases to air by chemical species for reporting year 2009. About 99% of the reported releases are hydrochloric acid mists, sulfuric acid mists (SOx), and hydrogen fluoride. Only about 1% of the reported releases are all the other chemicals listed in Table 1.5.

1.5 Mercury Releases

The comparisons made in the preceding sections are misleading in some ways. For example, there is hardly another industry sector that releases as much mercury into the environment as coal-fired power plants.

Mercury is a naturally occurring element that is found in air, water, and soil. It exists in several forms: elemental or metallic mercury, inorganic mercury compounds, and organic mercury compounds. Mercury is an element in the earth's crust. Pure mercury is a liquid metal, which alchemists referred to as quicksilver. The chemical readily volatizes.

Mercury is found in many rocks including coal. When coal is burned, mercury is released into the environment. Coal-burning power plants are generally recognized as the largest human-caused source of mercury emissions to the air in the United States, accounting for over 50% of all domestic human-caused mercury emissions.[2] The EPA has reported that about one-quarter of U.S. emissions from coal-burning power plants are deposited within the contiguous United States, and the remainder enters the global cycle. Burning hazardous wastes, producing chlorine, breaking mercury products, and spilling mercury, as well as the improper treatment and disposal of products or wastes containing mercury, can also release it into the environment.

In the United States, mercury compounds are still manufactured in small amounts for specialty uses, such as chemical and pharmaceutical applications. Larger quantities of these chemical compounds are generated as byproducts from pollution-control activities at gold mines or in various industrial wastes. Elemental mercury is processed in the United States from byproduct mercury compounds, and an unknown quantity of mercury compounds is imported into the United States for conversion to elemental mercury.

[2] 2005 National Emissions Inventory. <http://www.epa.gov/ttnchie1/net/2005inventory.html>

Table 1.5 Breakdown of Reported Releases of HAPs to Air for 2009 TRI Reporting for the 95 Plants Surveyed

Chemical	Pounds	Percent
Hydrochloric Acid (1995 and After "Acid Aerosols" Only)	2,306,042,888	75.71
Sulfuric Acid (1994 and After "Acid Aerosols" Only)	447,785,332	14.70
Hydrogen Fluoride	264,505,063	8.68
Ammonium Nitrate (Solution)	8,125,460	0.27
Barium Compounds	6,660,484	0.22
Selenium Compounds	2,361,680	0.08
Zinc Compounds	2,235,653	0.07
Vanadium Compounds	1,930,127	0.06
Manganese Compounds	1,539,084	0.05
Chromium Compounds (Except Chromite Ore Mined in the Transvaal Region)	881,189	0.03
Nickel Compounds	871,500	0.03
Lead Compounds	681,384	0.02
Arsenic Compounds	540,185	0.02
Styrene	423,110	0.01
Nitric Acid	385,827	0.01
Mercury Compounds	341,039	0.01
Formaldehyde	265,000	0.01
Cobalt Compounds	166,594	0.01
Chlorine	64,041	0.00
Antimony Compounds	46,313	0.00
Beryllium Compounds	34,265	0.00
Thallium Compounds	33,628	0.00
Acetophenone	33,000	0.00
Phenanthrene	26,000	0.00
N-Hexane	16,106	0.00
1,2,4-Trimethylbenzene	15,585	0.00
Naphthalene	10,899	0.00
Polycyclic Aromatic Compounds	7,461	0.00
Molybdenum Trioxide	3,995	0.00
Toluene	2,211	0.00
Xylene (Mixed Isomers)	1,630	0.00
Ethylene Glycol	1,510	0.00
Formic Acid	1,360	0.00
Polychlorinated Biphenyls	1,273	0.00
Dioxin and Dioxin-Like Compounds	1,066	0.00
Benzene	910	0.00
Ethylbenzene	451	0.00
Silver Compounds	79	0.00
Hexachlorobenzene	74	0.00
Benzo(g,h,i)Perylene	57	0.00
Fluorine	18	0.00
Diisocyanates	1	0.00
Total	3,046,043,531	100.00

Airborne mercury settles into water or onto land, where it is washed into water. Once it is deposited, certain microorganisms can change it into methyl mercury, which is a highly toxic form that bioaccumulates in fish, shellfish, and animals that eat fish. Fish and shellfish are the main sources of methyl mercury exposure to humans. Methyl mercury builds up more in some types of fish and shellfish than others. The levels of methyl mercury in fish and shellfish depend on what they eat, how long they live, and how high they are in the food chain.

The EPA works with the U.S. Food and Drug Administration (FDA) to issue advice to the public on how often they should eat certain types of commercially caught fish and shellfish.[3] In addition, the EPA releases an annual summary of information on locally issued fish advisories and safe-eating guidelines to the public.

In the U.S. EPA's Mercury Study Report to Congress (1997), the EPA estimated that 7% of women of childbearing age would have blood mercury concentrations greater than those equivalent to the Reference Dose (RfD). The estimate of 7% of women of childbearing age above the RfD was based on patterns of fish and shellfish consumption and methyl mercury concentrations present in fish and shellfish. Blood mercury analyses in the 1999–2000 National Health and Nutrition Examination Survey (1999–2000 NHANES) for 16-to-49 year old women showed that approximately 8% of women in the survey had blood mercury concentrations greater than 5.8 μg/L (which is a blood mercury level equivalent to the current RfD). Based on this prevalence for the overall U.S. population of women of reproductive age and the number of U.S. births each year, it is estimated that more than 300,000 newborns each year may have increased risk of learning disabilities associated with in utero exposure to methyl mercury.

Source: http://www.epa.gov/hg/exposure.htm

Another exposure to mercury is breathing mercury vapor. These exposures can occur when elemental mercury or products that contain elemental mercury break and release mercury to the air, particularly in warm or poorly ventilated indoor spaces.

Mercury exposure at high levels can harm the brain, heart, kidneys, lungs, and immune system of people of all ages. Research shows that most people's

[3] In 2004 the EPA and FDA issued the first-ever joint consumer advice about methyl mercury in fish and shellfish. This advice was for women who might become pregnant, women who are pregnant, nursing mothers, and young children. The advisory provides three recommendations for selecting and eating fish or shellfish to ensure that women and young children will receive the benefits of eating fish and shellfish and be confident that they have reduced their exposure to the harmful effects of methyl mercury. The EPA also hosts a web-based compilation of fish advisories issued by states, tribes, territories, and local governments.

fish consumption does not cause a health concern. However, it has been demonstrated that high levels of methyl mercury in the bloodstream of unborn babies and young children may harm the developing nervous system, making the child less able to think and learn.

Birds and mammals that eat fish are more exposed to mercury than other animals in water ecosystems. Similarly, predators that eat fish-eating animals may be highly exposed. At high levels of exposure, methylmercury's harmful effects on these animals include death, reduced reproduction, slower growth and development, and abnormal behavior.

As reported in Table 1.5, power plants emit a range of toxic metals. In addition to mercury emissions, coal-fired power plants emit antimony, beryllium, cadmium, cobalt, lead, manganese, nickel, arsenic, chromium, and selenium. Human exposure to these toxic pollutants can cause a range of health impacts such as cancer; damage to the liver, kidney, and the nervous and circulatory systems; respiratory effects, including a type of asthma specific to nickel; decreased lung function; and bronchitis.

The 2009 TRI reported mercury emissions for 456 electric-generating facilities in the United States (including Puerto Rico). Most of these plants are conventional coal-fired power plants; however, among them are some that co-fire waste coal and/or petroleum coke. These plants reported a combined 35.4 tons, or 72,778 pounds, of mercury air emissions released into the atmosphere in 2009. For comparative purposes, Table 1.6 tabulates the releases of mercury from the top 20 (least overall polluting plants in the TRI 2009 reporting) and the bottom 25 (i.e., the 25 most polluting plants in terms of overall air emissions). The comparison shows an order of magnitude difference between those facilities that employ modern pollution controls and those that do not. The reader must bear in mind that mercury bioaccumulates in the

Table 1.6 Comparison of Least and Most Polluting Plants Based on 2009 TRI

Facility	Owner	State	County	Hg (Lbs)
	Top 20—Least Polluting Plants			
Clover	Dominion	VA	Halifax	15.0
Craig	Tri-State Generation and Transmission	CO	Moffat	30.0
Huntington	Berkshire Hathaway—MidAmerican Energy	UT	Emery	25.8
San Juan	PNM Resources	NM	San Juan	47.8
Stuart	DPL	OH	Adams	94.0
Wansley	Southern Company	GA	Carroll	79.0
Brayton Point	Dominion	MA	Bristol	50.6

Table 1.6 Comparison of Least and Most Polluting Plants Based on 2009 TRI—*cont'd*

Facility	Owner	State	County	Hg (Lbs)
Gibson	Duke Energy	IN	Gibson	141.8
Bruce Mansfield	FirstEnergy	PA	Beaver	137.0
Cross	State of South Carolina	SC	Berkeley	114.4
St. Johns River/Northside	City of Jacksonville	FL	Duval	107.0
Cumberland	Tennessee Valley Authority	TN	Stewart	140.0
Winyah	State of South Carolina	SC	Georgetown	56.5
La Cygne	Great Plains Energy	KS	Linn	87.5
Belews Creek	Duke Energy	NC	Stokes	146.1
Hunter	Berkshire Hathaway—MidAmerican Energy	UT	Emery	96.2
Paradise	Tennessee Valley Authority	KY	Muhlenberg	170.0
Ghent	E.ON	KY	Carroll	139.9
Big Bend	TECO Energy	FL	Hillsborough	100.0
Roxboro	Progress Energy	NC	Person	83.0
			Average	93.1
Bottom 25—Most Polluting Plants				
Martin Lake	Luminant	TX	Rusk	1,566.0
Big Brown	Luminant	TX	Freestone	1,362.0
Labadie	Ameren	MO	Franklin	1,297.3
Miller	Southern Company	AL	Jefferson	1,205.7
Limestone	NRG Energy	TX	Limestone	1,077.0
Monticello	Luminant	TX	Titus	1,063.0
Scherer	Southern Company	GA	Monroe	888.5
Coal Creek	Great River Energy	ND	Mclean	861.8
Gavin	American Electric Power	OH	Gallia	852.0
Rockport	American Electric Power	IN	Spencer	852.0
Monroe	DTE Energy	MI	Monroe	848.2
W.A. Parish	NRG Energy	TX	Fort Bend	844.6
Sandow 4	Luminant	TX	Milam	809.0
Big Cajun 2	NRG Energy	LA	Pointe Coupee	798.0
Keystone	RRI Energy	PA	Armstrong	795.0
H.W. Pirkey	American Electric Power	TX	Harrison	791.0
Columbia (WI)	Alliant Energy	WI	Columbia	626.2
Sherburne County	Xcel Energy	MN	Sherburne	624.5
Milton R, Young	Minnkota Power Cooperative	ND	Oliver	587.0
Independence	Entergy	AR	Independence	582.0
Nebraska City	State of Nebraska	NE	Otoe	572.7
Conemaugh	RRI Energy	PA	Indiana	569.4
Laramie River	Basin Electric Power Cooperative	WY	Platte	560.0
Colstrip	PPL	MT	Rosebud	560.0
White Bluff	Entergy	AR	Jefferson	559.0
			Average	846.1

environment, and hence, the longer poorly performing plants are allowed to operate, the more long-term contamination issues generations to come will face.

The EPA has now proposed the first national standards for mercury, arsenic, and other toxic air pollution from coal-fired power plants. The proposed regulations would require many power plants to install commonly used pollution control technologies to decrease emissions of mercury, arsenic, chromium, nickel, and acid gases. The EPA alleges that the regulations will prevent up to 17,000 premature deaths and 11,000 heart attacks per year; additionally, the EPA states that the regulations will provide health benefits for children, preventing 120,000 cases of childhood asthma symptoms and about 11,000 fewer cases of acute bronchitis among children each year. The proposed regulations are also projected to avert more than 12,000 emergency room visits and hospital admissions and 850,000 fewer days of work missed due to illness. The EPA also notes that the regulations will support 31,000 short-term construction jobs and 9,000 long-term utility jobs. The EPA posits that the regulations will cost the industry around $10 billion (a figure in line with some industry estimates; other estimates are much higher) and provide benefits worth $100 billion. There seems to be some consensus that the rules will also result in the closing of a number of power plants for which it is not economic to come into compliance. It is expected that the power industry will institute litigation to challenge the validity of these new rules if adopted.

2 Solid Wastes

Chapter Outline

2.1 Overview

The U.S. Department of Energy defines waste coal as usable material that is a byproduct of previous coal processing operations. Waste coal is composed of mixed coal, soil, and rock (mine waste). Most waste coal is burned as is in unconventional fluidized-bed combustors. For some uses, waste coal may be partially cleaned by removing some extraneous noncombustible constituents. Examples of waste coal include fine coal, coal obtained from a refuse bank or slurry dam, anthracite culm, bituminous gob, and lignite waste. Waste coal is referred to as "culm" in the Eastern Pennsylvania anthracite fields and as "gob" or "boney" in the bituminous coal mining regions.

Before burning, coal is crushed and washed, creating wastewater. The wastewater contains a wide range of toxins including mercury, arsenic, beryllium, cadmium, nickel, selenium, other toxic heavy metals, and radionuclides.

Roughly 1.05 billion tons of coal are burned each year in the United States to generate electricity. Additionally, the more than 500 coal burning plants across the United States emit 1.9 billion tons of carbon dioxide each year and create 120 million tons of toxic waste. Coal combustion waste (CCW) constitutes the second largest waste stream after municipal solid waste in the United States.

Sulfur dioxide scrubbers that are used for air pollution controls also create toxic waste. The flue-gas desulfurization (FGD) process creates a wet solid residue containing calcium sulfite ($CaSO_3$) and calcium sulfate ($CaSO_4$). Dry

Handbook of Pollution Prevention and Cleaner Production Vol. 4. DOI: http://dx.doi.org/10.1016/B978-1-4377-7815-1.00002-3

material such as fly ash is added to stabilize the sludge for transport and landfill storage.

Coal waste is disposed of in landfills generally referred to as ponds or surface impoundments. The ponds are lined with compacted clay soil, a plastic sheet, or both. When rain filters through the toxic ash pits, the toxic metals are leached out and pushed downward by gravity toward the lining and the soil below. Because all liners eventually degrade, crack, or tear, landfills eventually leak and release their toxins into the groundwater. The Environmental Protection Agency has reported that a 10-acre landfill would leak 0.2 to 10 gallons per day, or between 730 and 36,500 gallons over a 10-year period. Table 2.1 lists a few of the larger ponds across the country that have been identified by the U.S. Environmental Protection Agency as high-risk impoundments either because of age or reported groundwater contamination problems.

2.2 Coal Combustion Byproducts

The term "coal combustion byproducts" (CCB) includes fly ash, bottom ash, boiler slag, and FGD (flue-gas desulfurization) material known as synthetic gypsum. An American Society for Testing and Materials (ASTM) subcommittee under Committee E-50 on Environmental Assessment addressed the question of standards and definitions of coal and CCB terms. The definitions for some of the pertinent terms are as follows:

- Fly ash is coal ash that exits in a combustion chamber in the flue gas and is captured by air pollution control equipment, such as electrostatic precipitators, baghouses, or wet scrubbers.
- Class C fly ash is fly ash that meets criteria for that class as defined in ASTM C618 for use in concrete.
- Class F fly ash is fly ash that meets criteria for that class as defined in ASTM C618 for use in concrete.
- Bottom ash consists of agglomerated ash particles—which are formed in pulverized-coal boilers—that are too large to be carried in the flue gases and adhere to the boiler walls or fall through open grates to an ash hopper at the bottom of the boiler.
- Boiler slag is molten ash collected at the base of the slag tap and cyclone boilers that is quenched with water and shatters into black, angular particles that have a smooth, glassy appearance.
- FGD is the process of removing gaseous sulfur dioxide (SO_2) from boiler exhaust gas. Primary types of FGD processors are wet scrubbers and dry scrubbers; sorbent injection is another primary process. SO_2 sorbents include lime, limestone, sodium-based compounds, and high-calcium fly ash.

Table 2.1 List of High-Risk Ash Ponds

Company	Facility Name	Unit Name	Town/State
Allegheny Energy	Pleasants Power Station	McElroy's Run Embankment	Willow Island, WV
American Electric Power	Big Sandy Plant	Fly Ash	Louisa, KY
American Electric Power	Cardinal Plant	Fly Ash Reservoir 2	Brilliant, OH
American Electric Power	Gavin Plant	Fly Ash Pond	Cheshire, OH
American Electric Power	Gavin Plant	Bottom Ash Pond	Cheshire, OH
American Electric Power	Amos Plant	Fly Ash Pond	St. Albans, WV
American Electric Power	Mitchell Plant	Fly Ash Pond	Moundsville, WV
American Electric Power	Muskingum River Plant	Unit 5 Bottom Ash Pond (Lower Fly Ash Pond)	Waterford, OH
American Electric Power	Muskingum River Plant	Upper Fly Ash Pond	Waterford, OH
American Electric Power	Muskingum River Plant	Middle Fly Ash Pond	Waterford, OH
American Electric Power	Philip Sporn Power Plant	Fly Ash Pond	New Haven, WV
American Electric Power	Tanners Creek Plant	Fly Ash Pond	Lawrenceburg, IN
Arizona Electric Power Cooperative	Apache Generating Station	Ash Pond 4	Cochise, AZ
Arizona Electric Power Cooperative	Apache Generating Station	Ash Pond 1	Cochise, AZ
Arizona Electric Power Cooperative	Apache Generating Station	Ash Pond 3	Cochise, AZ
Arizona Electric Power Cooperative	Apache Generating Station	Scrubber Pond 2	Cochise, AZ
Arizona Electric Power Cooperative	Apache Generating Station	Scrubber Pond 1	Cochise, AZ
Arizona Electric Power Cooperative	Apache Generating Station	Evaporation 1	Cochise, AZ
Arizona Electric Power Cooperative	Apache Generating Station	Ash Pond 2	Cochise, AZ
Arizona Public Service Company	Cholla Generating Station	Bottom Ash Pond	Joseph City, AZ
Arizona Public Service Company	Cholla Generating Station	Fly Ash Pond	Joseph City, AZ
Duke Energy	G.G. Allen Steam Plant	Active Ash Pond	Belmont, NC
Duke Energy	Belews Creek Steam Station	Active Ash Pond	Walnut Cove, NC
Duke Energy	Buck Steam Station	New Primary Pond	Spencer, NC
Duke Energy	Buck Steam Station	Secondary Pond	Spencer, NC
Duke Energy	Buck Steam Station	Primary Pond	Spencer, NC
Duke Energy	Dan River Steam Station	Secondary Pond	Eden, NC
Duke Energy	Dan River Steam Station	Primary Pond	Eden, NC
Duke Energy	Marshall Steam Station	Active Ash Pond	Terrell, NC

(Continued)

Table 2.1 List of High-Risk Ash Ponds—*cont'd*

Company	Facility Name	Unit Name	Town/State
Duke Energy	Riverbend Steam Station	Secondary Pond	Mount Holly, NC
Duke Energy	Riverbend Steam Station	Primary Pond	Mount Holly, NC
Dynegy Midwest Generation	Havana Power Station	East Ash Pond	Havana, IL
Dynegy Midwest Generation	Wood River Station	East Ash Pond (2 cells)	Alton, IL
FirstEnergy	Bruce Mansfield Power Station	Little Blue Run Dam	Shippingport, PA
Southern Company-owned Georgia Power	Branch Generating Plant	E	Milledgeville, GA
E.ON-owned Kentucky Utilities Company	E.W. Brown Generating Station	Auxiliary Pond	Harrodsburg, KY
E.ON-owned Kentucky Utilities Company	E.W. Brown Generating Station	Ash Pond	Harrodsburg, KY
E.ON-owned Kentucky Utilities Company	Ghent Generating Station	Gypsum Stacking Facility	Ghent, KY
E.ON-owned Kentucky Utilities Company	Ghent Generating Station	Ash Pond Basin 1	Ghent, KY
E.ON-owned Kentucky Utilities Company	Ghent Generating Station	Ash Pond Basin 2	Ghent, KY
E.ON-owned Louisville Gas & Electric Co.	Cane Run Station	Ash Pond	Louisville, KY
PPL Montana LLC	Colstrip Steam Plant	Units 1 & 2 Stage Evaporation Ponds (STEP)	Colstrip, MT
Progress Energy Carolinas Inc.	Asheville Plant	1982 Pond	Arden, NC
Progress Energy Carolinas Inc.	Asheville Plant	1964 Pond	Arden, NC

- FGD material is the product of an FGD process that typically uses a high-calcium sorbent, such as lime or limestone. Sodium-based sorbent and high-calcium fly ash are also used in some systems. The physical nature of these materials varies from a wet, thixotropic sludge to a dry, powdered material, depending on the process. FGD units remove SO_2 from flue gas but, in doing so, generate large quantities of synthetic gypsum (FGD material), which is a mixture of gypsum ($CaSO_4C_2H_2O$), calcium sulfite ($CaSO_3$), fly ash, and unreacted lime or limestone. A number of power plants convert the $CaSO_3$ to calcium sulfate ($CaSO_4$) by forced oxidation and take appropriate measures to reduce other impurities in the synthetic material and, thus, produce synthetic gypsum that meets or exceeds the specifications for

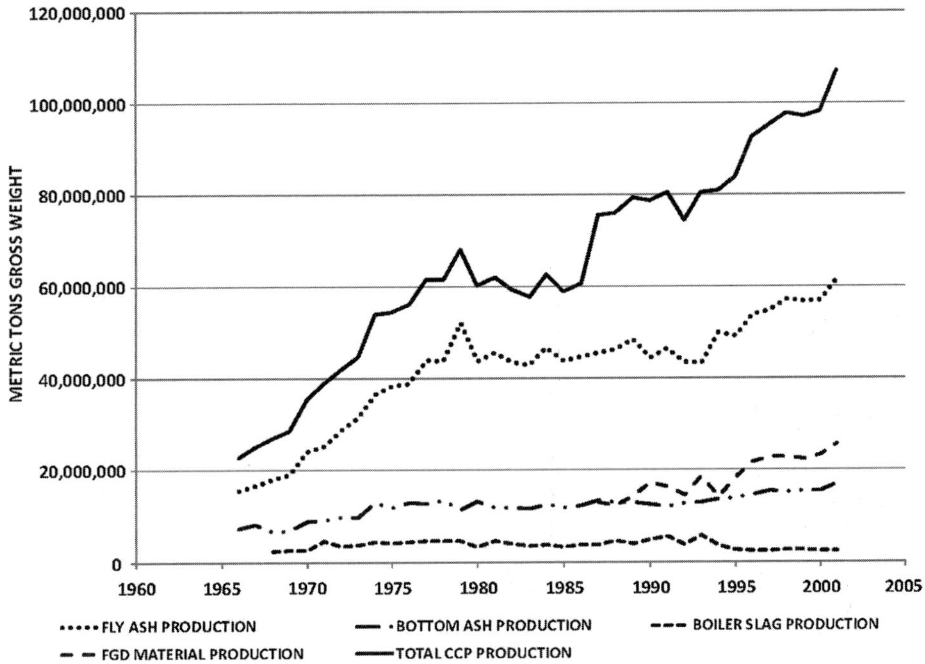

Figure 2.1 Reported CCB amounts generated over the years.
Source: *U.S. Geological Survey Minerals Yearbook 2001.*

> wallboard manufacture. Wallboard plants that have been constructed adjacent to such electric utilities use the FGD gypsum from those electric utilities. About 26 metric tons (MT) of FGD material were produced in 2001, and about 7.3 MT (28%) were used, mostly for wallboard manufacture. FGD issues affect, directly or indirectly, coal, gypsum, lime, limestone, and soda ash industries. Increased commercial use of FGD products represents an economic opportunity for high-sulfur coal producers and the sorbent industry. Synthetic gypsum competes directly with natural gypsum as raw material for wallboard and cement manufacture.

The value of CCBs is well established by research and commercial practice in the United States and abroad. Kalyoncu provides an overview of industry applications.[1]

The *U.S. Geological Survey Minerals Yearbook* publishes data based on industry surveys. The surveys conducted over the years account for about 60% of the power (highest survey response conducted in 2001). The last survey was conducted in 2001. Figures 2.1 and 2.2 were prepared using this information.

[1] R. Kalyoncu, "Coal Combustion Products," *U.S. Geological Survey Minerals Yearbook,* 2001.

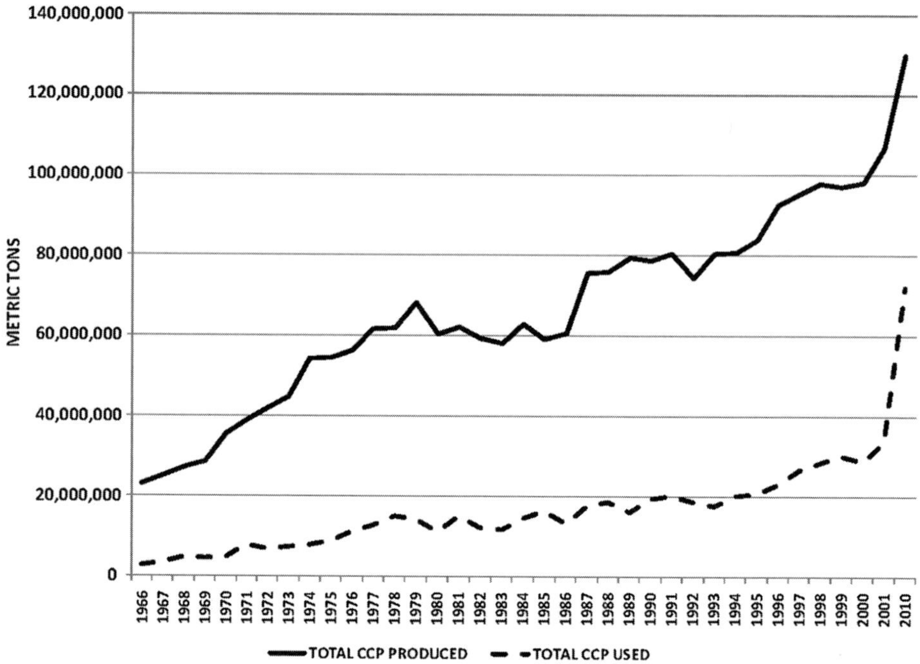

Figure 2.2 Reported amounts of CCBs used in industry applications.

Figure 2.1 shows the relative amounts of CCBs generated across the country based on projections made by the U.S. Geological Services analysis. Figures 2.2 and 2.3 show the amounts of CCBs used and the fraction of this waste reused in industry applications, respectively.

The U.S. Energy Information Administration (EIA) has published an estimate that nearly half of the electricity generated in the United States comes from coal, resulting in about 130 million tons of CCB as of 2010. One source reports that about 55%, or 72 million tons, is fly ash and that about half the fly ash generated in the United States is currently used for beneficial purposes, mostly in concrete.[2] This statistic is included as the 2010 data value in Figures 2.2 and 2.3. With the downturn in the economy and the lack of a peer-reviewed study to substantiate the claim, the reported estimate should not necessarily be considered accurate.

Figure 2.4 illustrates the industry general management practices in the handling of CCBs. Between 1966 and 2001, roughly 1.8 billion MTs of CCBs were generated. Of this amount, about 0.54 billion MTs was used in various

[2] William D. Palmer, Jr. "Fly Ash Threat," *Concrete Construction,* January 2010; http://www.concreteconstruction.net/concrete/fly-ash-threat.aspx.

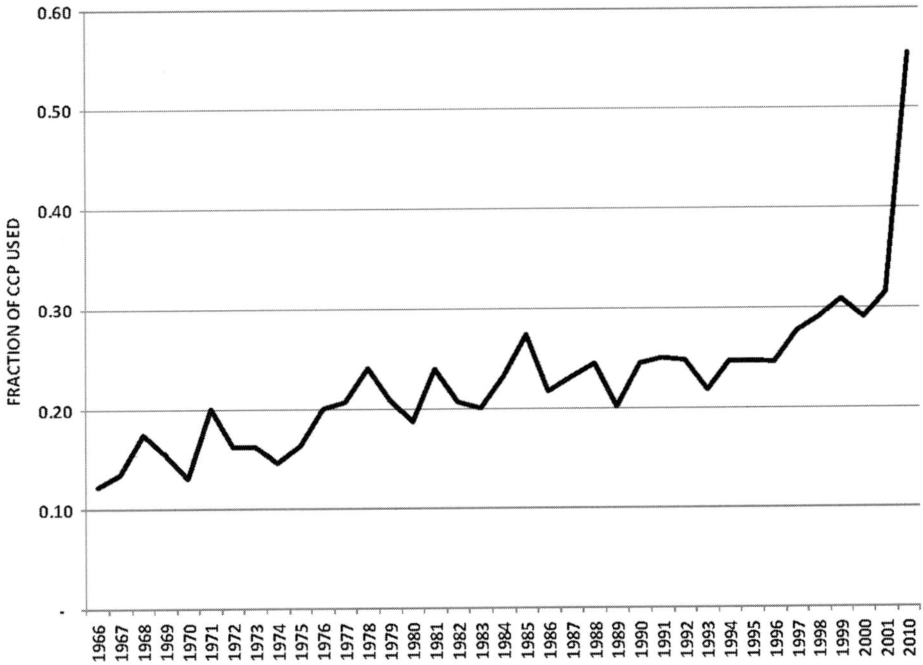

Figure 2.3 Fraction of CCB usage based on statistics reported by the U.S. Geological Survey.

industry applications (largely cement) and about 1.25 billion MTs sent to landfills or retained in ponds. Approximately 30% of the waste generated over this time period found use in value-added markets. Based on the 2010 reported value, a significant increase in use of this waste occurred in which 55% of the generated waste found its way into commercial markets.

2.3 Regulations

The Environmental Protection Agency does not regulate fly ash under Subtitle D of the Resource Conservation and Recovery Act (RCRA). The individual states have jurisdiction over its disposal, unlike RCRA Subtitle C hazardous wastes, which are regulated by the federal government. This approach has been in place since 1980 under the so-called Bevill exclusion, which excluded "solid waste from the extraction, beneficiation, and processing of ores and minerals" from regulation as hazardous waste under Subtitle C of RCRA. In 2000, the EPA determined that coal combustion products (CCBs), which include fly ash, bottom ash, and flue-gas desulfurization material, were exempt due to the desire to avoid placing "barriers on the beneficial uses of

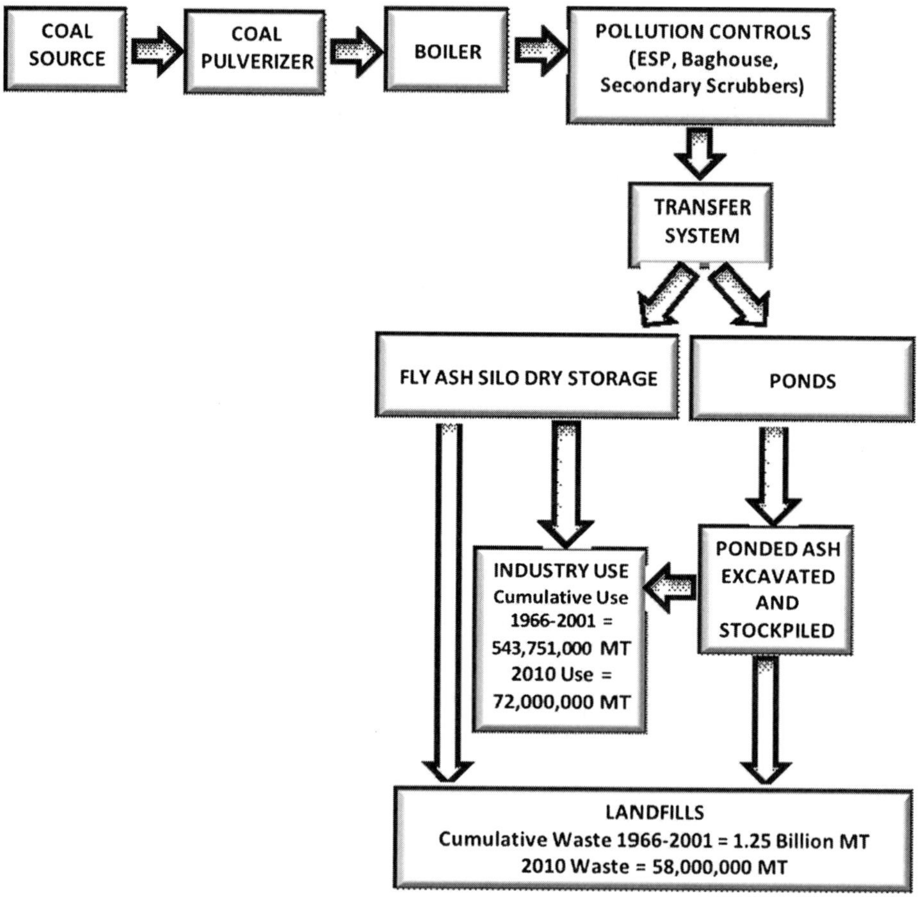

Figure 2.4 The fate of CCBs.

coal combustion waste so they can be used in applications that conserve natural resources and reduce disposal costs."[3]

Coal waste ponds contain billions of gallons of fly ash that generally include other coal waste containing toxic heavy metals. While the EPA considers these wastes a threat to water supplies and human health, as noted previously, they are not subject to federal regulation. A consequence of this is that there is lax monitoring of their impacts on the local environment.

The EPA reclassified fly ash from waste to a reusable material in the 1980s. The agency exempted ash from regulations for hazardous waste beginning in 1993. In 2001, the EPA stated that it wanted to set a national standard for

[3] http://www.fas.org/sgp/crs/misc/R40544.pdf

ponds or landfills that are used for the disposal of coal waste. However, the agency has yet to act. Consequently, coal ash ponds are subject to less regulation than municipal landfills.

In May 2009, an EPA representative announced at an energy industry conference that the agency is preparing regulations on how to handle ash from coal-fired power plants. At that time the EPA said coal ash may be reclassified as hazardous waste. Industry officials were vocal with objections, saying such a change would greatly increase disposal costs.

On December 17, 2009, the EPA announced it was postponing its findings on coal ash regulations. A final decision had been expected before the end of the year. The EPA attributed the delay to the complexity of the analysis the agency is currently finishing. In October 2009, the EPA sent the White House a Notice of Proposed Rulemaking for Toxic Coal Ash. President Obama's choice as the head of the White House Office of Management and Budget oversees such policies but as of March 2010 had not acted on the EPA's recommendation. The Office of Management and Budget has come under scrutiny for allowing the office to meet with coal industry representatives more than 20 times since October 2010. All such meetings took place behind closed doors and were not open to the public.

On March 4, 2010, the Alabama House introduced a bill that would allow Perry County, Alabama, to levy a $5 per ton fee on coal ash disposed at a privately owned landfill in the city of Uniontown. Alabama Rep. Ralph Howard of Greensboro, Alabama, introduced the bill. Currently, the Tennessee Valley Authority (TVA) is shipping coal sludge that breached the Kingston Fossil Plant in Tennessee. The TVA anticipates that it will ship approximately 3 million cubic feet of coal and ash to the landfill before the cleanup is completed. Revenue from the levy would be spent evenly between the towns of Uniontown and Marion. The total amount raised could be as much as $15 million.

2.4 Origins and General Properties of Fly Ash

Fly ash is a product of burning finely ground coal in a boiler to produce electricity. It is removed from the plant exhaust gases primarily by electrostatic precipitators (ESPs) or baghouses, and secondarily by scrubber systems. In pulverized coal boilers, fly ash is produced during char burnout from the melting of inorganic mineral matter in the coal by a series of processes involving initial mineral fragmentation followed by coalescence on the char surface. The molten ash particles, entrained in the combustion gases after char fragmentation, undergo quenching, forming in part spherical, glass-like particles as they are swept away from the flame region.

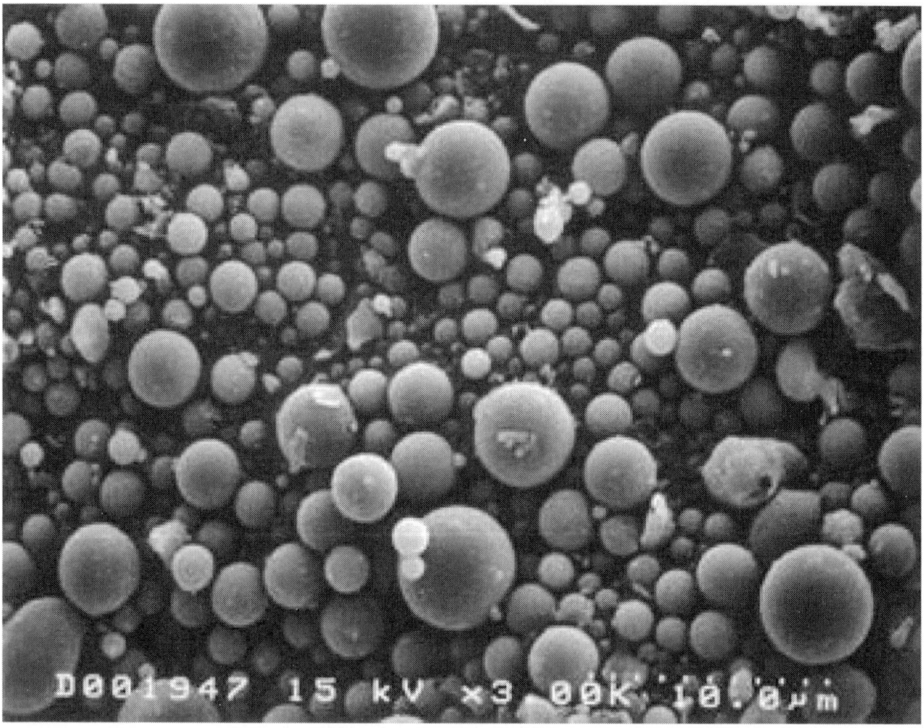

Figure 2.5 Photograph of fly ash at 2000× magnification.
Source: U.S. Department of Transportation website, http://www.fhwa.dot.gov/pavement/recycling/fach01.cfm

A number of investigators have reported microanalysis of ash collected in flue-gas cleaning equipment in which the fly ash consists primarily of spherical particles of impure aluminosilicate glass. The particle size varies from submicrometer to > 100 micrometers (μm). Figure 2.5 shows a photograph of fly ash at 2000× magnification.

The U.S. Department of Transportation describes fly ashes as "finely divided residue resulting from the combustion of ground or powdered coal. They are generally finer than cement and consist mainly of glassy-spherical particles as well as residues of hematite and magnetite, char, and some crystalline phases formed during cooling."[4]

Two major classes of fly ash are specified in ASTM C 618 on the basis of their chemical composition resulting from the type of coal burned; these are designated Class F and Class C.

[4] http://www.fhwa.dot.gov/infrastructure/materialsgrp/flyash.htm

- Class F is fly ash normally produced from burning anthracite or bituminous coal.
- Class C is normally produced from the burning of subbituminous coal and lignite.

Class C fly ash usually has cementitious properties in addition to pozzolanic properties due to free lime, whereas Class F is rarely cementitious when mixed with water alone.

The U.S. Department of Transportation's website also states that fly ash which is produced at base-loaded electric-generating plants is "usually very uniform." Base-loaded plants are defined as those plants which operate continuously. The only exception to uniformity is in the start-up and the shutdown of these plants. Contamination may occur from using other fuels to start the plant, and inconsistencies in carbon content occur until the plant reaches full operating efficiency. The ash produced from the start-up and shutdown must be separated from what is produced when the plant is running efficiently. In addition, when sources of coal are changed, it is necessary to separate the two types of fly ashes.

In contrast, peak load plants are subjected to frequent start-up and shutdown cycles. Because of this, these plants may not produce consistently uniform fly ash.

There are wide differences in characteristics within each fly ash class. Despite the reference in ASTM C 618 to the classes of coal from which Class F and Class C fly ashes are derived, there is no requirement that a given class of fly ash must come from a specific type of coal. For example, Class F ash can be derived from coals that are not bituminous, and bituminous coals can produce ash that is not Class F.

Fly ash ranges in color from light tan to dark gray depending on its chemical and mineral constituents. Tan and light colors are associated with high lime content. A brownish color is associated with the iron content. A dark gray to black color is typically attributed to an elevated unburned carbon content. Fly ash color is generally consistent for each power plant and coal source. Its makeup consists in part of silt-sized and clay-sized glassy spheres. This gives fly ash a consistency somewhat like talcum powder.

Ghosal and Self reported physical characterization of six ashes from coals representative of those burned in U.S. power plants.[5] Centrifugal separation was used to classify the ashes into six density categories in the range < 1.6 to > 3.2 g cm^{-3}. The size distributions of all density classes were determined in the range 1–200 μm. For all six ashes, the median diameter initially decreased and then increased with increasing particle density.

[5] S. Ghosal and A. Self, "Particle Size-Density Relation and Cenosphere Content of Coal Fly Ash," *Fuel*, 1995. 74(4), 522–529.

Using centrifugal separation, the investigators estimated the mass fraction of cenospheres (i.e., particles with trapped interior "bubbles") in the ashes to vary from < 5 to > 95 wt%. The cenosphere content was found to be uncorrelated with coal rank but was positively correlated with the total mineral content of the coal. The median diameters of the cenospheric fractions were found to be two to three times those of the noncenospheric (solid) fractions. Density-size data were used to determine the Fe_2O_3 distribution in the ashes. The density classification of the six U.S. coal ashes showed > 80 wt % of four of the samples to lie between 2.0 and 2.8 g-cm^{-3}, whereas two-thirds of a high-calcium ash had a density range of 2.8–3.2 g-cm^{-3}. The cenosphere fraction of the ash was defined as that of density < 2g-cm^{-3}. Among the samples tested, the cenosphere fraction was found to vary from < 5 wt% to > 95 wt%. Additionally, the median diameter of the cenosphere fraction was found to be two- to three-fold larger than the noncenosphere fraction. The investigators also noted that for some of the ashes, a considerable fraction of the cenospheres contained only small bubbles, and particle densities were 2.0–2.2 g-cm^{-3}.

Another observation reported in the study is that as much as 50% of the Fe_2O_3 in the ash is concentrated in large spherical particles that are typically more than 20 μm in size, which were found to be relatively few in number and some of which are nonglassy. The cenosphere content of an ash was found to be directly related to the total ash content of the coal.

The ash that isn't used, though, ends up in landfills or containment ponds. To handle the ash, some electric utilities mix it with water and pump the slurry into containment ponds. Unfortunately, these ponds have not always been well designed or maintained, which is what led to the Tennessee disaster and the current political situation, and why the EPA's administrator, Lisa Jackson, pledged during her confirmation hearing in January 2009 to come up with new disposal regulations for CCBs by the end of that year. The EPA doesn't trust the states to properly regulate fly ash but has no legal way to regulate it federally other than to designate the material as hazardous under Subtitle C, even though the toxicity of the material does not qualify it as hazardous.

2.5 Flue-Gas Desulfurization Material

Flue-gas desulfurization (FGD) material is a product of a process typically used for reducing SO_2 emissions from the exhaust gas system of a coal-fired boiler. The physical nature of these materials varies from a wet sludge to a dry, powdered material depending on the process. The wet sludge from a lime-based reagent wet scrubbing process is predominantly calcium sulfite. The wet product from limestone-based reagent wet scrubbing processes is predominantly calcium sulfate. The dry material from dry scrubbers that is captured in a baghouse consists of a mixture of sulfites and sulfates. This powdered material is referred to as dry FGD ash, dry FGD material, or lime spray dryer ash. FGD gypsum consists of small, fine particles.

Calcium sulfite FGD material can be used as an embankment and road base material. Calcium sulfate FGD material, once it has been dewatered, can be used in wallboard manufacturing and in place of gypsum for the production of cement. The largest single market for FGD material is in wallboard manufacturing.

2.6 Toxic Heavy Metals Found in Fly Ashes

The EPA and others have reported a broad composition of heavy metals in fly and bottom ashes. Table 2.2 reports ranges and average compositions in parts per million (ppm).

According to the *U.S. Geological Survey Minerals Yearbook 2001,* the cumulative amount of CCB deposited in landfills and surface ponds between 1966 and 2001 is 1.8 billion MT. Using the average metals composition reported in Table 2.2, the amount of toxic metals contained in this waste is 16,547,096 MT. These metals represent both a significant environmental threat as well as an enormous potential reclaimed value. Using average metal value prices reported in 2010–2011, the value of the metals contained in this waste is valued at more than $14 trillion. See Table 2.3. The monetary value constitutes a sizable incentive to develop technologies for extracting these metals as opposed to landfilling and stockpiling, which pose risks to the environment.

To further explore the incentives, Table 2.4 lists 100 U.S. coal-burning plants and the amount of ash reported to have been placed in surface impoundments for one reporting year. The total amount of CCB placed in ponds amounts to nearly 115 million pounds. Using average compositions for metals in fly ash, about 1,057,357 pounds of heavy metals are contained in the waste. Applying the same cost values for metals reported previously, the value of the metals is more than $415 million. Table 2.5 provides a breakdown of the estimated value of the metals for each facility.

2.7 Beneficial Ash Characteristics and Markets

The ash generated from coal combustion is produced when pulverized coal ($70\% < 75$ μm) is injected into the furnace with preheated air, and rapid

Table 2.2 Composition of Heavy Metals in PPM Reported in Fly Ashes from Coal Combustion

		Reference [1]		**Reference [2]**		**Mean Literature Values**	
Toxin	**Symbol**	**Min**	**Max**	**Min**	**Max or Single Value**	**Min**	**Max**
Arsenic	As	17	510	2.3	6,300	9.7	3,405
Antimony	Sb	3	14			1.5	7
Beryllium	Be				5		2.5
Barium	Ba	590	7,000		806	295	3,903
Boron	Bo			10	618	5	309
Cadmium	Cd	0.3	1.8	0.7	130	0.5	65.9
Chromium	Cr	66	210	10	1,000	38	605
Chromium VI	Cr -VI				90		45
Cobalt	Co	16	66	7	520	11.5	293
Copper	Cu				112		56
Lead	Pb	24	120	3.1	5,000	13.6	2,560
Manganese	Mn			58	3,000	29	1,500
Mercury	Hg		1.5		1		1.3
Molybdenum	Mo	6.9	77			3.5	38.5
Nickel	Ni			6.3	4,300	3.2	2,150
Selenium	Se	1.1	210	0.2	134	0.7	172
Strontium	Sr				775		387.5
Thallium	Tl	0.7	13		9	0.4	11
Vanadium	Va				252		126
Zinc	Zn			10	3,500	5	1,750

[1]"Characterization of Coal Combustion Residues from Electric Utilities: Leaching and Characterization Data," EPA-600R-09/151, December 2009 (data taken from Table 13, p. 183, average reported values for 34 fly ash samples).
[2]"Managing Coal Combustion Residues in Mines," Committee on Mine Placement of Coal Combustion Wastes, National Research Council of the National Academies, 2006.

combustion occurs. Approximately 80% of the ash is carried out by the combustion gases and is fly ash, and the remaining 20% drops to the bottom of the furnace (referred to as bottom ash). Fly ash is fine powder with a mean particle size of roughly 50 μm. The fly ash is transported from the combustion chamber by the exhaust gas and is recovered by electrostatic precipitation or by other methods.

Fly ash is a heat-activated silica rich material. The silica is the main beneficial element when fly ash is added to Portland cement concrete. Fly ash is a pozzolan, which is a siliceous or a siliceous and aluminous material that, in the presence of water, will combine with an activator (lime, Portland cement, or kiln dust) to produce a cementitious material. As already noted, there are two types of fly ash as defined by ASTM:

Class F has $SiO_2 + Al_2O_3 + Fe_2O_3 \exists 70\%$.

Table 2.3 Estimates of Quantities and Value of Heavy Metals in Cumulative Ash Waste (1996–2001)

Toxin	Metric Tons through 2001	Pounds	Cost ($) per Lb	Value (Millions $)
Total	16,547,096	36,486,346,998		14,337,888.62
Barium	3,770,704	8,314,403,359	249.70	2,076,106.52
Arsenic	3,067,088	6,762,929,354	1,452.80	9,825,183.77
Lead	2,311,600	5,097,077,838	45.40	231,407.33
Nickel	1,933,991	4,264,449,164	0.60	2,558.67
Zinc	1,576,366	3,475,888,017	0.82	2,840.50
Manganese	1,373,370	3,028,280,785	1.27	3,849.55
Strontium	696,116	1,534,936,304	454.00	696,861.08
Chromium	577,552	1,273,501,991	1.27	1,618.88
Boron	282,039	621,896,773	2,270.00	1,411,705.68
Cobalt	273,506	603,081,425	95.34	57,497.78
Vanadium	226,350	499,101,869	12.26	6,117.99
Selenium	155,077	341,944,197	24.06	8,227.86
Copper	100,600	221,823,053	3.09	685.43
Cadmium	59,641	131,509,381	12.00	1,578.11
Molybdenum	37,680	83,084,617	30.00	2,492.54
Thallium	10,204	22,499,195	217.92	4,903.02
Antimony	7,635	16,834,785	2.00	33.63
Beryllium	4,491	9,902,815	422.22	4,181.17
Mercury	2,246	4,951,407	7.90	39.11

Class C has $SiO_2 + Al_2O_3 + Fe_2O_3 \exists 50\%$.

CSA A23.5 defines three types of fly ash:

Type F has $CaO < 8\%$.
Type CI has CaO between 8% and 20%.
Type CH has $CaO > 20\%$.

Bottom ash is mainly highly abrasive sand-sized particles that normally are sluiced away by a high-pressure water stream. Bottom ash can be used as an abrasive for sand blasting and as an aggregate for cement.

During combustion, sulfur dioxide (SO_2) is formed from sulfur in the coal. If allowed to escape, it reacts with water to form sulfuric acid (H_2SO_4), the prime component in acid rain. An effective method of collecting this gas (after the fly ash has been recovered) is to scrub exiting flue gas with a lime or limestone slurry to form gypsum ($CaSO_4.2H_2O$). This finds a ready market in wallboard and cement manufacture, and in agriculture.

Table 2.4 Surface Impoundment Releases and Estimated Amounts of Heavy Metals

Rank	Facility	Corporate Owner	City	State	2006 Surface Impoundment Releases (lbs)	Total Lbs of Metals
1	Stanton Energy Center	Orlando Utilities Commission	Orlando	FL	8,423,056	77,586
2	Sherburne County Generating Plant	Xcel Energy	Becker	MN	4,721,862	43,494
3	Coal Creek Station	Great River Energy	Underwood	ND	4,372,709	40,277
4	Scherer Steam Electric Plant	Georgia Power/ Southern Company	Juliette	GA	4,114,502	37,899
5	Detroit Edison Monroe Power Plant	DTE Energy	Monroe	MI	4,110,859	37,866
6	Gibson Generating Station	Duke Energy	Owensville	IN	3,030,524	27,914
7	Gorgas Steam Plant	Alabama Power/ Southern Company	Parrish	AL	2,888,290	26,604
8	Cholla Power Plant	Arizona Public Service Company	Joseph City	AZ	2,863,427	26,375
9	Wansley Steam Plant	Georgia Power/ Southern Company	Roopville	GA	2,673,672	24,627
10	Ghent Generating Station	E.ON US	Ghent	KY	2,664,501	24,543
11	J.M. Stuart Station	Dayton Power & Light, Duke, AEP	Manchester	OH	2,456,637	22,628
12	Harllee Branch Generating Plant	Georgia Power/ Southern Company	Milledgeville	GA	2,433,945	22,419
13	Barry Steam Plant	Alabama Power/ Southern Company	Bucks	AL	2,350,349	21,649
14	Gaston Steam Plant	Alabama Power/ Southern Company	Wilsonville	AL	2,306,006	21,241
15	Miller Steam Plant	Alabama Power/ Southern Company	Quinton	AL	2,160,349	19,899

Table 2.4 Surface Impoundment Releases and Estimated Amounts of Heavy Metals—*cont'd*

Rank	Facility	Corporate Owner	City	State	2006 Surface Impoundment Releases (lbs)	Total Lbs of Metals
16	La Cygne Generating Station	Great Plains Energy	Lacygne	KS	2,127,000	19,592
17	Gallatin Fossil Plant	Tennessee Valley Authority	Gallatin	TN	2,093,068	19,279
18	Boswell Energy Center	Minnesota Power	Cohasset	MN	2,009,628	18,511
19	Leland Olds Station	Basin Electric Power Cooperative	Stanton	ND	1,937,821	17,849
20	Widows Creek Fossil Plant	Tennessee Valley Authority	Stevenson	AL	1,864,177	17,171
21	Paradise Fossil Plant	Tennessee Valley Authority	Drakesboro	KY	1,765,148	16,259
22	Labadie Power Station	AmerenUE	Labadie	MO	1,740,882	16,035
23	Kingston Fossil Plant	Tennessee Valley Authority	Harriman	TN	1,738,437	16,013
24	Cardinal Plant	American Electric Power	Brilliant	OH	1,707,225	15,725
25	Bowen Steam Plant	Georgia Power /Southern Company	Cartersville	GA	1,684,118	15,513
26	Pearl Station	Soyland Power Cooperative	Pearl	IL	1,661,744	15,306
27	New Madrid Power Plant	Associated Electric Cooperative	Marston	MO	1,514,440	13,950
28	Kammer and Mitchell Plants	American Electric Power	Moundsville	WV	1,372,687	12,644
29	Kyger Creek Station	Ohio Valley Electric Corp.	Cheshire	OH	1,356,475	12,495
30	Greene County Steam Plant	Alabama Power/ Southern Company	Forkland	AL	1,343,973	12,379
31	Baldwin Energy Station	Dynegy	Baldwin	IL	1,324,467	12,200

(*Continued*)

Table 2.4 Surface Impoundment Releases and Estimated Amounts of Heavy Metals—*cont'd*

Rank	Facility	Corporate Owner	City	State	2006 Surface Impoundment Releases (lbs)	Total Lbs of Metals
32	Rush Island Power Station	AmerenUE	Festus	MO	1,307,769	12,046
33	Karn and Weadock Generating Plants	Consumers Energy	Essexville	MI	1,171,382	10,790
34	Cayuga Generating Station	Duke Energy	Cayuga	IL	1,154,623	10,635
35	Council Bluffs Energy Center	MidAmerican Energy	Council Bluffs	IA	1,092,320	10,061
36	Chesterfield Power Station	Dominion	Chester	VA	1,088,260	10,024
37	Milton R. Young Station	Minnkota Power Cooperative	Center	ND	1,036,290	9,545
38	Wabash River Generating Station	Duke Energy	W. Terre Haute	IN	951,610	8,765
39	A.B. Brown Generating Station	Vectren	Mount Vernon	IL	944,944	8,704
40	Big Sandy Plant	American Electric Power	Louisa	KY	915,079	8,429
41	Amos Plant	American Electric Power	Winfield	WV	864,024	7,959
42	Big Cajun II	NRG Energy	New Roads	LA	860,640	7,927
43	Hammond Steam Generating Station	Georgia Power/ Southern Company	Rome	GA	849,068	7,821
44	Tanners Creek Plant	American Electric Power	Lawrenceburg	IN	819,840	7,552
45	Muskingum River Plant	American Electric Power	Beverly	OH	791,757	7,293
46	Mayo Generating Plant	Progress Energy	Roxboro	NC	786,128	7,241
47	Killen Generating Station	Dayton Power & Light, Duke Energy	Manchester	OH	715,435	6,590
48	Roxboro Steam Plant	Progress Energy	Semora	NC	698,290	6,432

Table 2.4 Surface Impoundment Releases and Estimated Amounts of Heavy Metals—*cont'd*

Rank	Facility	Corporate Owner	City	State	2006 Surface Impoundment Releases (lbs)	Total Lbs of Metals
49	Trimble County Generating Station	E.ON US	Bedford	KY	637,434	5,871
50	E.W. Brown Generating Station	E.ON US	Harrodsburg	KY	637,230	5,870
51	George Neal Station North	MidAmerican Energy	Sergeant Bluff	IA	612,005	5,637
52	Clifty Creek Station	Ohio Valley Electric Corp.	Madison	IN	590,808	5,442
53	Welsh Power Plant	American Electric Power	Pittsburg	TX	562,064	5,177
54	Coleto Creek Power Station	International Power	Fannin	TX	550,623	5,072
55	L.V. Sutton Electric Plant	Progress Energy	Wilmington	NC	548,210	5,050
56	Laramie River Station	Basin Electric Power Cooperative	Wheatland	WY	541,970	4,992
57	Lansing Smith Generating Plant	Gulf Power/ Southern Company	Southport	FL	520,282	4,792
58	Naughton Power Plant	PacifiCorp/ MidAmerican Energy	Kemmerer	WY	517,966	4,771
59	Meramec Power Plant	AmerenUE	St. Louis	MO	481,318	4,433
60	Shawnee Fossil Plant	Tennessee Valley Authority	West Paducah	KY	467,616	4,307
61	Brayton Point Station	Dominion	Somerset	MA	464,254	4,276
62	Duck Creek Station	Ameren	Canton	IL	462,272	4,258
63	Twin Oaks Power Station	OptimEnergy	Bremond	TX	449,002	4,136
64	Conesville Power Plant	American Electric Power	Conesville	OH	447,846	4,125
65	G.G. Allen Steam Plant	Duke Energy	Belmont	NC	439,208	4,046
66	Montrose Station	Great Plains Energy	Clinton	MO	422,100	3,888

(*Continued*)

Table 2.4 Surface Impoundment Releases and Estimated Amounts of Heavy Metals—*cont'd*

Rank	Facility	Corporate Owner	City	State	2006 Surface Impoundment Releases (lbs)	Total Lbs of Metals
67	Allen Fossil Plant	Tennessee Valley Authority	Memphis	TN	416,705	3,838
68	Cliffside Plant	Duke Energy	Mooresboro	NC	413,459	3,808
69	Asheville Plant	Progress Energy	Arden	NC	411,793	3,793
70	Meredosia Power Station	Ameren	Meredosia	IL	398,106	3,667
71	Louisa Generating Station	MidAmerican Energy	Muscatine	IA	382,063	3,519
72	Asbury Generating Station	Empire District Electric Co.	Asbury	MO	381,186	3,511
73	H.W. Pirkey Power Plant	American Electric Power	Hallsville	TX	380,111	3,501
74	Yates Steam Generating Plant	Georgia Power/ Southern Company	Newnan	GA	376,610	3,469
75	Joppa Steam Plant	Ameren	Joppa	IL	366,675	3,377
76	Havana Power Station	Ameren	Havana	IL	360,772	3,323
77	Apache Generating Station	Arizona Electric Power Cooperative	Cochise	AZ	360,465	3,320
78	Canadys Station	SCE&G/ SCANA	Canadys	SC	357,897	3,297
79	Lee Steam Plant	Progress Energy	Goldsboro	NC	356,078	3,280
80	Kincaid Generating Station	Dominion	Kincaid	IL	355,108	3,271
81	Cape Fear Steam Plant	Progress Energy	Moncure	NC	334,076	3,077
82	Intermountain Power Station	Intermountain Power Service Corp.	Delta	UT	333,589	3,073
83	Frank Ratts Generating Station	Hoosier Energy	Petersburg	IN	330,014	3,040
84	McDonough/ Atkinson Steam Plant	Georgia Power/ Southern Company	Smyrna	GA	318,051	2,930

Table 2.4 Surface Impoundment Releases and Estimated Amounts of Heavy Metals—*cont'd*

Rank	Facility	Corporate Owner	City	State	2006 Surface Impoundment Releases (lbs)	Total Lbs of Metals
85	Petersburg Generating Station	AES	Petersburg	IN	309,961	2,855
86	Dolet Hills Power Station	Cleco	Mansfield	LA	291,208	2,682
87	Rockport Plant	American Electric Power	Rockport	IN	281,995	2,597
88	Buck Steam Station	Duke Energy	Spencer	NC	279,190	2,572
89	Hugo Plant	Western Farmers Electric Cooperative	Hugo	OK	275,203	2,535
90	Wood River Station	Dynegy	Alton	IL	267,066	2,460
91	Gallagher Generating Station	Duke Energy	New Albany	IN	260,183	2,397
92	Oklaunion Power Station	American Electric Power	Vernon	TX	254,652	2,346
93	Gadsden Steam Plant	Alabama Power/ Southern Company	Gadsden	AL	249,740	2,300
94	Iatan Generating Station	Great Plains Energy	Weston	MO	240,245	2,213
95	Sioux Power Plant	AmerenUE	West Alton	MO	226,193	2,083
96	Flint Creek Power Plant	American Electric Power	Gentry	AR	221,456	2,040
97	Riverton Power Plant	Empire District Electric Company	Riverton	KS	212,688	1,959
98	Spurlock Power Station	East Kentucky Power Cooperative	Maysville	KY	196,954	1,814
99	Jeffrey Energy Center	Westar Energy	St. Marys	KS	190,417	1,754
100	W.S. Lee Steam Station	Duke Energy	Pelzer	SC	190,030	1,750

Table 2.5 Potential Value of Metals in Ash Waste by Facility

Rank	Facility	Total Lbs of Metals	Total $
1	Stanton Energy Center	77,586	30,488,513
2	Sherburne County Generating Plant	43,494	17,091,487
3	Coal Creek Station	40,277	15,827,675
4	Scherer Steam Electric Plant	37,899	14,893,056
5	Detroit Edison Monroe Power Plant	37,866	14,879,870
6	Gibson Generating Station	27,914	10,969,436
7	Gorgas Steam Plant	26,604	10,454,598
8	Cholla Power Plant	26,375	10,364,603
9	Wansley Steam Plant	24,627	9,677,756
10	Ghent Generating Station	24,543	9,644,560
11	J.M. Stuart Station	22,628	8,892,166
12	Harllee Branch Generating Plant	22,419	8,810,028
13	Barry Steam Plant	21,649	8,507,440
14	Gaston Steam Plant	21,241	8,346,934
15	Miller Steam Plant	19,899	7,819,707
16	La Cygne Generating Station	19,592	7,698,995
17	Gallatin Fossil Plant	19,279	7,576,173
18	Boswell Energy Center	18,511	7,274,150
19	Leland Olds Station	17,849	7,014,233
20	Widows Creek Fossil Plant	17,171	6,747,668
21	Paradise Fossil Plant	16,259	6,389,218
22	Labadie Power Station	16,035	6,301,383
23	Kingston Fossil Plant	16,013	6,292,533
24	Cardinal Plant	15,725	6,179,557
25	Bowen Steam Plant	15,513	6,095,917
26	Pearl Station	15,306	6,014,931
27	New Madrid Power Plant	13,950	5,481,742
28	Kammer and Mitchell Plants	12,644	4,968,646
29	Kyger Creek Station	12,495	4,909,964
30	Greene County Steam Plant	12,379	4,864,712
31	Baldwin Energy Station	12,200	4,794,107
32	Rush Island Power Station	12,046	4,733,666
33	Karn and Weadock Generating Plants	10,790	4,239,993
34	Cayuga Generating Station	10,635	4,179,331
35	Council Bluffs Energy Center	10,061	3,953,816
36	Chesterfield Power Station	10,024	3,939,120
37	Milton R. Young Station	9,545	3,751,007
38	Wabash River Generating Station	8,765	3,444,495
39	A.B. Brown Generating Station	8,704	3,420,366
40	Big Sandy Plant	8,429	3,312,265
41	Amos Plant	7,959	3,127,464
42	Big Cajun II	7,927	3,115,215
43	Hammond Steam Generating Station	7,821	3,073,329
44	Tanners Creek Plant	7,552	2,967,534

Table 2.5 Potential Value of Metals in Ash Waste by Facility—*cont'd*

Rank	Facility	Total Lbs of Metals	Total $
45	Muskingum River Plant	7,293	2,865,883
46	Mayo Generating Plant	7,241	2,845,508
47	Killen Generating Station	6,590	2,589,624
48	Roxboro Steam Plant	6,432	2,527,565
49	Trimble County Generating Station	5,871	2,307,288
50	E.W. Brown Generating Station	5,870	2,306,549
51	George Neal Station North	5,637	2,215,244
52	Clifty Creek Station	5,442	2,138,518
53	Welsh Power Plant	5,177	2,034,475
54	Coleto Creek Power Station	5,072	1,993,062
55	L.V. Sutton Electric Plant	5,050	1,984,328
56	Laramie River Station	4,992	1,961,742
57	Lansing Smith Generating Plant	4,792	1,883,239
58	Naughton Power Plant	4,771	1,874,856
59	Meramec Power Plant	4,433	1,742,203
60	Shawnee Fossil Plant	4,307	1,692,606
61	Brayton Point Station	4,276	1,680,437
62	Duck Creek Station	4,258	1,673,263
63	Twin Oaks Power Station	4,136	1,625,230
64	Conesville Power Plant	4,125	1,621,046
65	G.G. Allen Steam Plant	4,046	1,589,779
66	Montrose Station	3,888	1,527,854
67	Allen Fossil Plant	3,838	1,508,326
68	Cliffside Plant	3,808	1,496,577
69	Asheville Plant	3,793	1,490,546
70	Meredosia Power Station	3,667	1,441,004
71	Louisa Generating Station	3,519	1,382,934
72	Asbury Generating Station	3,511	1,379,760
73	H.W. Pirkey Power Plant	3,501	1,375,869
74	Yates Steam Generating Plant	3,469	1,363,196
75	Joppa Steam Plant	3,377	1,327,235
76	Havana Power Station	3,323	1,305,868
77	Apache Generating Station	3,320	1,304,757
78	Canadys Station	3,297	1,295,462
79	Lee Steam Plant	3,280	1,288,878
80	Kincaid Generating Station	3,271	1,285,367
81	Cape Fear Steam Plant	3,077	1,209,238
82	Intermountain Power Station	3,073	1,207,475
83	Frank Ratts Generating Station	3,040	1,194,535
84	McDonough/Atkinson Steam Plant	2,930	1,151,233
85	Petersburg Generating Station	2,855	1,121,950
86	Dolet Hills Power Station	2,682	1,054,071
87	Rockport Plant	2,597	1,020,723

(*Continued*)

Table 2.5 Potential Value of Metals in Ash Waste by Facility—*cont'd*

Rank	Facility	Total Lbs of Metals	Total $
88	Buck Steam Station	2,572	1,010,570
89	Hugo Plant	2,535	996,138
90	Wood River Station	2,460	966,685
91	Gallagher Generating Station	2,397	941,771
92	Oklaunion Power Station	2,346	921,751
93	Gadsden Steam Plant	2,300	903,971
94	Iatan Generating Station	2,213	869,603
95	Sioux Power Plant	2,083	818,739
96	Flint Creek Power Plant	2,040	801,593
97	Riverton Power Plant	1,959	769,856
98	Spurlock Power Station	1,814	712,905
99	Jeffrey Energy Center	1,754	689,243
100	W.S. Lee Steam Station	1,750	687,842
Totals		**1,057,357**	**415,505,331**

Some utilities use petroleum (Pet) coke as a fuel feedstock with the coal. This Pet coke can make up 5–30% of the fuel. The ash produced is similar to the ash from coal alone. Pet coke is more difficult to burn than coal, and the principal difference in the ash is a slightly higher concentration of unburned carbon. If necessary, the carbon can be removed by ash beneficiation processes.

Most coals contain trace amounts of valuable minerals, and some coals that are rich in a particular mineral generate an enriched combustion product that might warrant recovery. Recovery usually involves conventional mining technology such as foam flotation or high-temperature gravity separation.

Fly ash use in Portland cement concrete is well established. Calcium-rich cementitious binders served the Greeks and Romans well where only strength was required. When durability was of concern, they found that silica in finely divided form was essential. Their source of silica was ground volcanic ash. Its effectiveness can be seen in the harbor at Cosa on the west coast of Italy, which was built about 273 BC with silica from Mount Vesuvius. Today we know that the proper proportions of fly ash in concrete can provide the following benefits: improve rheological properties, making it easier to pump, place, and finish concrete; increase the strength; reduce permeability to water; increase the resistance to chloride ion diffusion; decrease the corrosion rate of embedded steel; increase the resistance to sulphate attack; increase the resistance to acid attack; suppress deleterious expansion due to alkali–silica reaction (ASR); and minimize the risk of delayed ettringite formation (DEF).

But it is also known that the performance of fly ash concrete is influenced by the following:

- The type of fly ash (physical and chemical properties)
- The level of replacement used
- The nature of the Portland cement (e.g., alkali content)
- The quality of the concrete
- The curing, because the slower reaction of many fly ashes increases the sensitivity of the concrete to proper curing methods

A number of factors influence the setting time of concrete. Concrete temperature and the chemical composition of the cement are the most important. Increasing fly ash content usually increases both initial and final set time. At room temperature, each 10% increase of cement replacement with fly ash increases set time by about 5 to 20 minutes. At higher temperatures, the effect is less pronounced. At lower temperatures, concrete setting times can be greatly delayed and should be determined for the materials used. Some high-calcium fly ashes (i.e., ASTM Class C or CSA Type CH) lead to shorter setting times.

Most fly ash particles are solid spheres with some small amount being hollow cenospheres. The spheres act as pseudo ball bearings that increase both the consistency (slump) and workability of the concrete. Their effectiveness increases in direct proportion to their surface area. Normally, concrete mixtures with fly ash will require less water per cubic meter for a given slump than a mixture without fly ash. Table 2.6 summarizes the attributes achieved from the use of fly ash in cement formulations.

Fly ash has other beneficial uses. High-calcium ash can be used to counteract acidification. As such, it may be used as a soil conditioner. Because fly ash is highly alkaline, it can be spread on grassland to reduce the effect of acid rain. However, some fly ash may be deficient in important elements necessary for plant growth and/or may have an excess of others. Before fly ash is used for agricultural purposes, it is important to examine the fly ash composition as well as evaluate the soil under consideration to make sure that they are compatible and suitable for the intended purpose.

Calcium sulfate from flue-gas desulfurization (FGD) scrubbers can also be used to advantage on some crops such as potatoes and peanuts. Products from other FGD processes (e.g., lime spray dryers) can be used to increase calcium and sulfur values in acidic soils. Gypsum is used in agriculture for three principal purposes: as a source of calcium and sulfur as plant nutrients; as a chemical decompactant in heavy clay soils; and to restore ionic balance in soils with salt buildup from irrigation.

Fly ash and bottom ash can also be used for highway embankments provided optimum moisture is maintained and adequate compaction is provided. The source can come from an operating power plant, or reclaimed

Table 2.6 Benefits of Fly Ash

Performance Attribute	Description
Increased Workability	For mixtures with the same cementitious content, with and without fly ash, the one with fly ash will generally be noticeably more workable for a given consistency. Consequently, the coarse aggregate content usually can be increased when fly ash is used to replace cement.
Placing	The improved rheological properties associated with fly ash make the concrete easier to pump, consolidate, and finish.
Reduced Bleeding	The use of fly ash reduces the rate and amount of bleeding in fresh concrete. As a result, care must be taken to minimize evaporation of the mixture water after placing to avoid cracking due to plastic shrinkage.
Reduced Heat of Hydration	The exothermic reaction between cement and water can be reduced by replacing some of the cement with fly ash. Generally, the early-age heat generation of a cement fly ash mixture is 30% less than that of an equivalent mass of Portland cement. When high volume fly ash concrete is used (e.g., 56% fly ash with a superplasticizer), a substantial reduction in maximum temperature results, frequently enabling large sections to be cast without exceeding a maximum temperature of 40°C. With all concrete, attention must be paid to the cooling rate so that temperature differentials between the surface and interior do not exceed 20°C, so as to prevent surface cracking.
Increased Long-Term Strength	Although concrete mixtures containing fly ash tend to gain strength at a slower rate than concrete without fly ash, the long-term strength (90 days and after) is usually higher. The enhanced workability and increased consistency achieved when fly ash is used in high cementitious content mixtures allow a very significant reduction in water content. This reduction, combined with a superplasticizer, enables very high-strength concrete mixtures to be produced, typical of what is used for high-rise buildings. High-calcium fly ashes (ASTM Class C and CSA Type CH) will show a more rapid strength gain at early ages than concrete made with a lower calcium fly ash (i.e., Class or Type F).
Increased Volume Stability	Concrete achieves its volume stability mainly from the large quantity of fine and coarse aggregate used. These aggregates have a modifying or restraining effect on the volume changes that cement paste undergoes. Hydrating cement paste shrinks upon loss of water and tends to crack under the influence of an applied load. Fortunately, the much stiffer aggregate minimizes this volume change. Volume change in concrete containing fly ash is slightly less than that in concrete without fly ash when the concretes are compared at equal strengths.

Table 2.6 Benefits of Fly Ash—*cont'd*

Performance Attribute	Description
Reduced Permeability	During the hydration process, large spaces between the cement grains are filled with hydration products. With time, these hydration products become more dense as the pozzolanic effect of fly ash in the concrete mixture creates even more impermeable hydration products. Also the use of a pozzolan like fly ash tends to decrease the formation of weak boundary layers between the cement paste matrix and the normally impermeable aggregates. Although initially more permeable than a concrete made without fly ash after several years, a concrete with 25% fly ash can have a coefficient of permeability at least one order of magnitude less than a concrete without fly ash. This leads to enhanced durability because aggressive agents cannot attack the concrete from within but are restricted to the concrete surface.
Increased Resistance to Chloride Penetration	The penetration of chloride ions (from seawater or deicing salts) into concrete leads to corrosion of embedded steel reinforcement. The use of fly ash increases the resistance of concrete to chloride ingress, extending the service life of reinforced concrete exposed to a chloride environment. The increased resistance to chloride is demonstrated by the chloride profiles shown, which were established for plain Portland cement concrete and fly ash concrete in the same sea wall after 30 years of exposure.
Increased Resistance to Sulfates	Sulfates in soil or groundwater can attack the cement hydrates leading to deterioration of the concrete. The use of low-calcium fly ash (ASTM Class F or CSA Type F) can increase the sulfate resistance of concrete. High-calcium fly ashes are less effective in this role.
Increased Resistance to Alkali–Silica Reaction	Alkali–silica reaction (ASR) is a reaction between the alkali from Portland cement and certain siliceous aggregates. ASR can lead to internal expansion and cracking of concrete. The use of low-calcium fly ash (ASTM Class F or CSA Type F) reduces the potential for ASR in concrete. Higher calcium fly ashes are less effective in this role.
Resistance to Freezing and Thawing	It is well established that concrete will be resistant to cyclic freezing and thawing provided that • The aggregate is frost-resistant • Sufficient strength is attained prior to first freezing (> 5 MPa) • Sufficient strength is attained prior to cyclic freezing and thawing (> 20 MPa) • An adequate air-void system is present These conditions are equally applicable to concrete containing fly ash.

(*Continued*)

Table 2.6 Benefits of Fly Ash—*cont'd*

Performance Attribute	Description
Resistance to Deicer Salt Scaling	The destructive forces associated with cyclic freezing and thawing of concrete are exacerbated in the presence of deicing salts. A number of laboratory studies have indicated inferior scaling resistance of concrete containing levels of fly ash in excess of 25%–30% when subjected to cycles of freezing and thawing in the presence of deicing chemicals such as sodium chloride and calcium chloride. However, accelerated salt scaling tests in the laboratory do not provide a reliable indication of field performance, and concrete with fairly high levels of fly ash (e.g., 56%) have been used in various applications (e.g., sidewalks) exposed to deicer salts and freezing conditions with satisfactory performance. When fly ash concrete is exposed to deicing salts, great care should be taken to ensure the water-cementing materials ratio is kept to a minimum (preferably $W/CM < 0.40$), that the concrete is adequately air-entrained, and that proper finishing and curing procedures are followed.

ash can come from a lagoon or stockpiled ash. The product from an operating plant normally can be delivered with close limitations on moisture content. Moisture content of fly ash from a lagoon or from a stockpile can vary considerably depending on its location within the lagoon or stockpile.

In road applications, fly ash may serve as a low-density road base. Class F ash usually is hauled to the construction site in covered dump trucks with the water content adjusted by spraying water on the loading conveyor belt or in the case of lagoon ash by blending the ash with dryer silo ash. The ash is spread in lifts of 150 to 300 mm thick. The lift is then compacted to obtain the required in-place density. Depending on the ash quality, leachate is normally not a problem with compacted ash that is appropriately graded to impede infiltration when a properly seeded soil cover is used to control erosion. Before proceeding, one should check the local environmental regulations to see that proposed work meets with the regulations.

Compacted fly ash has reasonable mechanical behavior and compaction characteristics similar to silt. Where it can wick water from a shallow groundwater table, it also can share some of the difficulties of silt, such as being susceptible to frost heave as a result of ice lenses. Fly ash should not be used as a fill without an impermeable cover to prevent infiltration, below the groundwater table, or where a drainage layer is not used at the bottom of the fill to prevent water wicking.

Stabilized road bases can be produced by mixing fly ash and a calcium-rich material with aggregates. The calcium-rich material can be Portland cement, lime, or kiln dust. Class C fly ash frequently has sufficient calcium to serve both functions. Then it need only be mixed with the aggregates and water followed by compaction to achieve optimum density to generate the required in-place strength. A typical mixture will contain 8%–14% fly ash and 3%–8% lime by weight. Type I Portland cement can be used to accelerate early strength gain.

Other beneficial uses include construction markets and as an extender and fill in polymer compounding applications.

Fly ash is being used in the production of some bricks. It can improve the plasticity of the clay so that plant production rates can be increased and because the carbon in the fly ash reduces the fuel to fire the bricks. The addition of fly ash acts similarly to grog (pulverized reject brick) in that it reduces the shrinkage of the brick that takes place during the firing process. This latter aspect is of particular advantage when making hollow blocks, roofing tiles, and clay flower pots.

Fly ash has also found use in cellular concrete. Cellular concrete is made by mechanically incorporating foam into a cement paste matrix or by incorporating a gas-forming agent such as aluminum powder. In the latter case, the aluminum powder reacts with calcium hydroxide, liberating hydrogen gas. The hydrogen generates numerous well-dispersed noninterconnected small bubbles from 0.2 to 2 mm in diameter that cause the material to more than double in volume. After about 2 hours, the material has gained sufficient strength so that it can be cut into individual panels or blocks, and these building components are then placed in an autoclave for 6–12 hours. The chemical reactions are greatly speeded up at the high temperature and pressure and at the end of the cycle have a strength of up to 8 MPa. The product then has sufficient strength and volume stability to be used as masonry units and wall panels. Because of the vesicular nature of the product, it has a density of 800 kg/m^3 or less and has a thermal conductivity much less than that of regular concrete. About 30% of the binder can be fly ash, and it can be relatively coarse because the elevated temperature at high alkalinity within the autoclave causes rapid dissolution of the particles. A high carbon content tends to destabilize the air voids. A good fly ash for cellular concrete should have a high silica and low iron and carbon content.

Fly ash has found use as a flowable fill. Emergency repair of broken water mains requires excavation and re-installment of the roadway as quickly as possible, and the resulting repair should not result in settlement of the fill that will require repeated patching of the roadway. A lean slurry of cement, fly ash, and sand, plus chemical admixtures has been developed that produces a fill that can be discharged directly into an excavation without the need of vibration. This fill will not shrink or settle and at the same time is of sufficiently low strength that it can be excavated easily should the pipes need to be repaired or replaced in the future. Normally, flowable fill is used below ground, where freeze and thaw resistance is not a requirement. Fly ash content can be as high as 80% of the cementitious materia, and in the case of backfilling tunnels and conduits, high addition rates of superplasticizers are used.

Fly ash has found use in grouts and mortars. It is effective in grouts and mortars where the material has to be pumped and where mechanical or manual consolidation is not possible. The principal benefits are reduced bleeding, improved flowability and pumpability, reduced long-term permeability, increased long-term strength and durability, and enhanced economy. Many buildings have been built over abandoned mine workings where the voided ground is relatively close to the surface. Filling these caverns or tunnels with very lean cement mixtures can be economically attractive when up to 9% of the fill is composed of fly ash.

Fly ash is used as a filler for plastics and polymer compounding. When fly ash from a power plant is disposed of hydraulically in a lagoon, a small fraction of the fly ash floats. The floaters are hollow spheres about 50–200 μm in diameter and have a density that is about 70% of that of water. They have a wall thickness of about 10% of their radius and have a higher silica content and a lower calcium content than normal fly ash. As a highly effective inert filler, they have found a ready market for use in paints, plastics, refractory tiles, and the matrix of fibrous composites. Because of the limited supply and the expense of gathering and drying these cenospheres, their cost is quite high compared to other fly ash.

Fly ash can be used in plastics to reduce the cost and to provide dimensional stability to the end products, particularly for high-temperature applications. The spherical nature of fly ash allows greater addition rates than is possible with other additives where a high viscosity needs to be maintained during the injection process.

2.8 Radioactive Constituents

In addition to heavy metals, fly ash contains radioactive constituents. An article that was published in *Scientific American* reported the following findings.[6] The naturally occurring radioactive elements uranium and thorium

[6] M. Hvistendahl,, "Coal Ash Is More Radioactive than Nuclear Waste," *Scientific American*, December 13, 2007, http://www.scientificamerican.com/article.cfm?id=coal-ash-is-more-radioactive-than-nuclear-waste

exist in coal as trace elements. However, when coal is burned, leaving fly ash as a coal combustion byproduct, the uranium and thorium are concentrated at up to 10 times their original levels. As pointed out in the article, fly ash uranium can leach into the soil and ground and surface waters surrounding a coal plant, affecting cropland (thus entering the food chain), and/or become accessible through groundwater wells. Also, communities living within a "stack shadow"—the area within a half- to one-mile (0.8- to 1.6-kilometer) radius of a coal plant's smokestacks—might then ingest small amounts of radiation. Fly ash is also disposed of in landfills and abandoned mines and quarries, posing a potential risk to communities living around those areas.

The *Scientific American* article cites a 1978 publication that appeared in *Science*, in which J. P. McBride at Oak Ridge National Laboratory (ORNL) and his colleagues reported on the uranium and thorium content of fly ash from coal-fired power plants in Tennessee and Alabama. To answer the question of just how harmful leaching could be, the investigators estimated radiation exposure around the coal plants and compared it with exposure levels around boiling-water reactor and pressurized-water nuclear power plants. They reported estimated radiation doses ingested by people living near the coal plants to be equal to or higher than doses for people living around the nuclear facilities. At one extreme, the scientists estimated fly ash radiation in individuals' bones at around 18 millirems (thousandths of a rem, a unit for measuring doses of ionizing radiation) a year. Doses for the two nuclear plants, by contrast, ranged from between 3 and 6 millirems for the same period. In situations in which all food was grown in the area, radiation doses were 50%–200% higher around the coal plants.

3 Old and New Generation Technologies

Chapter Outline

3.1 Overview

Coal-fired power-producing plants fall into five categories: stoker-fired, pulverized coal (PC), cyclone-fired, fluidized-bed combustion (FBC), and coal gasification (IGCC). In the United States, pulverized coal is the coal-firing configuration predominantly used at existing electric utility power plants. It is also most frequently selected for new coal-fired projects.

Fluidized-bed combustion and coal gasification are actually old technologies that are being reinvented as newer technologies and being considered as alternatives to building new PC-fired electric-generating units. Cyclone- and stoker-firing technologies are considered outdated technologies that are generally not taken into account when building new coal-fired plants. There are, however, some existing cyclone- and stoker-fired units still in operation.

This chapter serves as a primer for these technologies.

3.2 Stoker-Fired Coal Combustion

Stoker-fired coal combustion technology dates back to the late nineteenth century. The coal is crushed and burned on a grate. Heated air passes upward through openings in the grate. Stokers are classified according to the way coal

Handbook of Pollution Prevention and Cleaner Production Vol. 4. DOI: http://dx.doi.org/10.1016/B978-1-4377-7815-1.00003-5

is fed to the grate—as underfeed stokers, overfeed stokers, and spreader stokers.

A spreader stoker uses a flipping mechanism to throw the coal into the furnace above the grate. The fine coal particles burn in suspension while heavier coal lumps fall to the grate and burn in a fuel bed.

In the underfeed configuration, the coal is fed by pushing the coal underneath the burning fuel bed.

In the traveling grate arrangement, coal is fed by gravity onto a moving grate and leveled by a stationary bar at the furnace entrance.

Stoker-firing coal combustion is an obsolete technology because the other, newer coal combustion technologies provide superior coal combustion efficiency, applicability, and other advantages. There are a few small stoker-fired power-generating facilities in service in the United States, but as these units are retired, no new stoker-fired coal systems are expected to be built. The majority of new stoker-fired boiler capacity is likely to occur at municipal solid waste combustor facilities and facilities burning solid biomass. These older systems are well suited for clean biomass fuel applications on a small scale.

3.3 Pulverized-Coal Combustion

When coal is pulverized into a very fine powder, the combustion process is more efficient. With a PC-fired unit, the coal is first pulverized in a mill to the consistency of talcum powder, which means that at least 70% of the particles will pass through a 200-mesh sieve. The pulverized coal is generally entrained in the primary combustion air before being blown through the burners into the combustion chamber, where it is fired in suspension. PC-fired boilers are classified by the firing position of the burners, either as wall-fired or tangential-fired.

PC units are predominantly used at existing U.S. electric-generating units. Roughly 92% of the total coal consumed by U.S. power plants is burned on these types of systems. They are considered the preferred coal-firing design of choice for new large coal-fired plants that are greater than 400 MWe.

There are basically two configurations: wall-fired and tangential-fired. Wall-fired systems are composed of an array of burners that fire into the furnace horizontally and can be positioned on one wall or opposing walls, depending on the specific furnace design.

Tangential-fired (also called corner-fired) units are composed of multiple burners that are positioned in opposite corners of the furnace, producing a fireball that moves in a cyclonic motion and expands to fill the volume of the furnace.

The U.S. Environmental Protection Agency provides a simplified process diagram of these units that is reproduced in Figure 3.1. The schematic shows

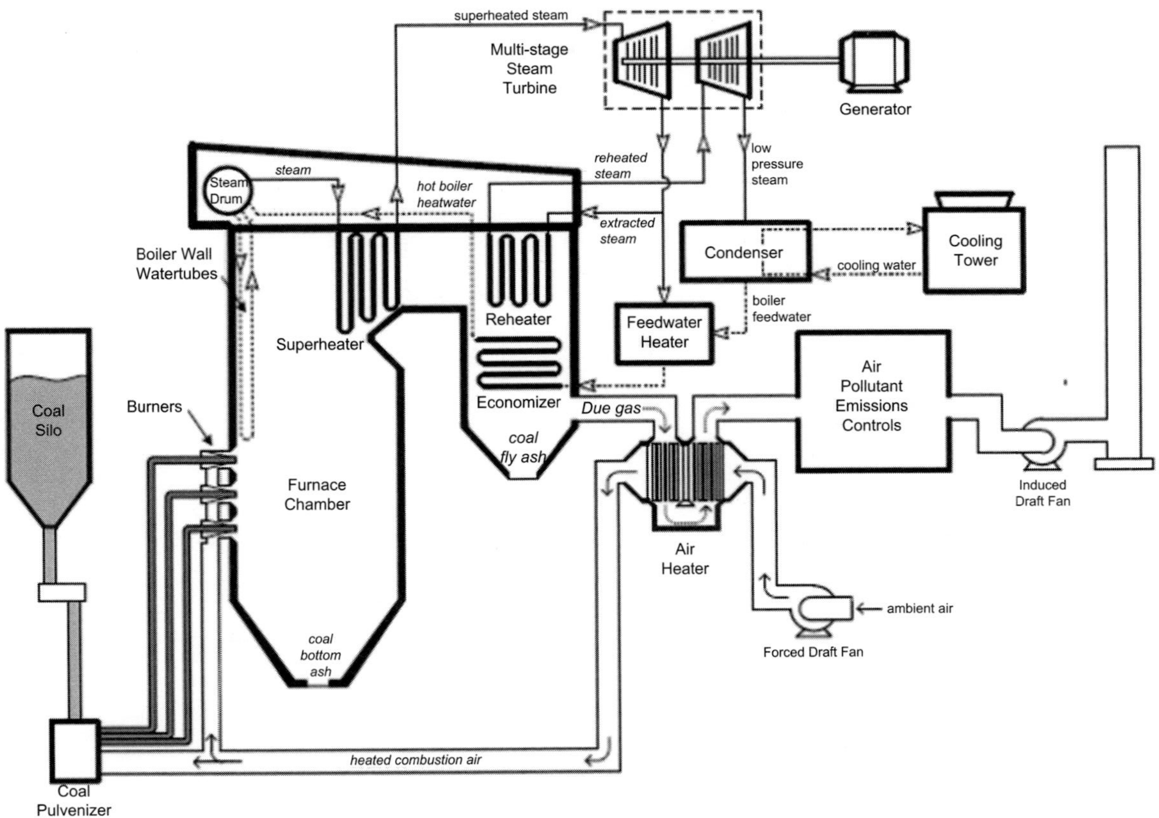

Figure 3.1 Schematic of PC-fired system using subcritical boiler.
Source: *Available and Emerging Technologies for Reducing Greenhouse Gas Emissions from Coal-Fired Electric Generating Units,* U.S. Environmental Protection Agency, Research Triangle Park, North Carolina, October 2010.

the major components of a PC-fired boiler using subcritical steam conditions. The pulverized coal is ignited and burned in the section of the boiler called the "furnace chamber" (also known as the "firebox"). Ambient air blown into the furnace chamber provides the oxygen required for combustion. The walls of the furnace chamber are lined with vertical tubes containing the feedwater. Heat transfer from the hot combustion gases in the furnace boils the water in the tubes to produce the high-temperature, high-pressure steam. The steam is separated from boiler water in a steam drum and sent to the steam turbine. The remaining water in the drum re-enters the boiler for further conversion to steam. The hot combustion products are vented from the furnace in a gas stream called collectively flue gas.

3.4 Cyclone Coal Combustion

Cyclone coal combustion technology requires less preprocessing of the coal and allows for the burning of lower-rank coals with higher moisture and ash contents. Cyclone boilers use burner design and placement (i.e., several water-cooled horizontal burners) to produce high-temperature flames that circulate in a cyclonic pattern.

Coal is crushed to a 4-mesh size and then fed tangentially with primary air to a horizontal cylindrical combustion chamber. The small coal particles are burned in suspension, while the larger particles are forced against the outer wall. The high temperatures developed in the relatively small boiler volume, combined with the low fusion temperature of the coal ash, cause the ash to form a molten slag, which is drained from the bottom of the boiler through a slag tap opening.

Cyclone units have high nitrogen oxide (NOx) emission rates. Most of the boiler systems of this configuration were placed into service in the early 1980s and before. The U.S. EPA reports that about 6% of total coal consumed by U.S. power plants passes through these units. This technology is designed to burn coals with low-ash fusion temperatures that are difficult to burn in PC boilers, whereby the majority of the ash is retained in the form of a molten slag. These designs are approaching extinction, and likely no new ones will be built in the coming years.

3.5 Fluidized-Bed Combustion

A fluidized bed is a mixture of fluid and solid particulate matter that exhibits fluid-like properties. The upper surface of the bed is relatively horizontal, which is analogous to hydrostatic behavior. The bed can be considered to be an inhomogeneous mixture of fluid and solid that can be represented by a single bulk density. The system behaves very much like a low-viscosity fluid when considered in a macroscopic scale. For example, an object with a higher density than the bed will sink, whereas an object with a lower density than the bed will float; thus, the bed can be considered to exhibit the fluid behavior expected of Archimedes' principle. The "density" (the solid volume fraction of the suspension) of the bed can be altered by changing the fluid fraction, and as such, objects with different densities comparative to the bed can, by altering either the fluid or solid fraction, be caused to sink or float.

In fluidized beds, the contact of the solid particles with the fluidization medium (gas) is greatly enhanced when compared to packed beds. This behavior in fluidized combustion beds enables high thermal transport inside the system and good heat transfer between the bed and its container. Similarly to the good heat transfer, which enables thermal uniformity analogous to that of a well-mixed gas, the bed can have a significant heat-capacity while maintaining a homogeneous temperature field.

In a fluidized-bed combustor (FBC), combustion occurs when coal and a sorbent, such as limestone, are suspended through the action of primary combustion air distributed below the combustor floor. The gas cushion between the solids allows the particles to move freely, giving the bed its liquid-like characteristic.

FBCs operate as atmospheric or pressurized boilers. There are two fluidized-bed design configurations: a bubbling fluidized bed and a circulating fluidized bed (CFB).

A CFB boiler is generally described as displaying a greater degree of fuel flexibility over a PC-fired boiler. These units are capable of combusting any rank of coal (including coal refuse), petroleum coke, and biomass without significant hardware modifications.

The combustion temperature of an FBC boiler (1,500°F to 1,650°F) is significantly lower than a PC-fired boiler (2,450°F to 2,750°F), which results in lower NOx formation and the ability to capture sulfur dioxide (SO_2) with limestone injection in the furnace. Although the combustion temperature of an FBC boiler is low, the circulation of hot particles provides efficient heat transfer to the furnace walls, thereby allowing for longer residence times for carbon combustion and limestone reaction. This results in good combustion efficiencies, comparable to PC-fired boilers.

Figure 3.2, taken from the U.S. EPA's publication, shows a simplified schematic of the components of a CFB boiler for an electric-generating plant. In this scheme, calcium in the sorbent combines with SO_2 gas to form calcium sulfite and sulfate solids, and solids exit the combustion chamber and flow into a hot cyclone. The cyclone separates the solids from the gases, and the solids

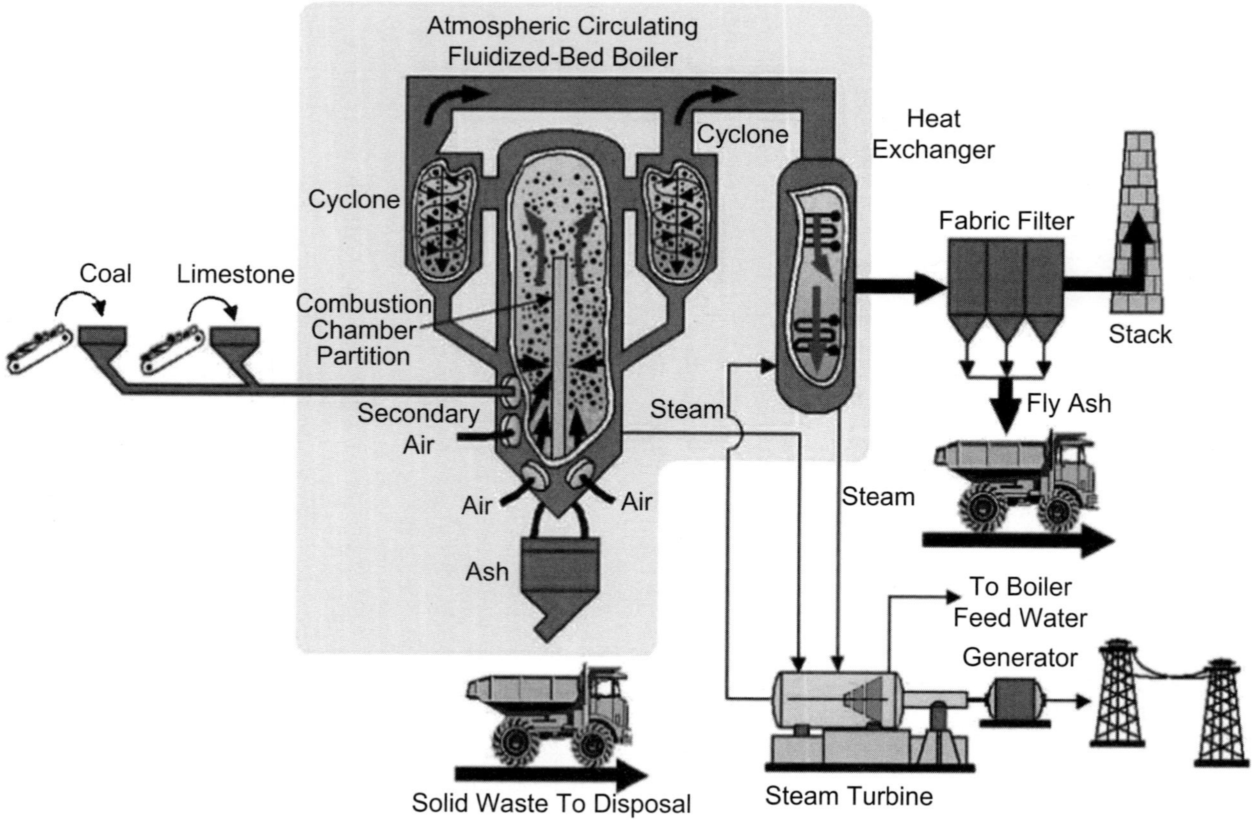

Figure 3.2 Schematic of atmospheric circulating fluidized-bed (CFB) boiler power plant.
Source: *Available and Emerging Technologies for Reducing Greenhouse Gas Emissions from Coal-Fired Electric Generating Units,* U.S. Environmental Protection Agency, Research Triangle Park, North Carolina, October 2010.

are recycled for combustor temperature control. Heat in the flue gas exiting the hot cyclone is recovered in a series of heat recovery sections of the boiler to produce steam. The superheated steam leaving the boiler then enters the steam turbine, which powers a generator to produce electricity. CFB boilers can be used with either subcritical or supercritical steam cycles.

The capacity of CFB subcritical boilers ranges from 25 to 350 MWe. The following are examples of commercial systems that are in operation:

- Two 300 MWe CFB subcritical boilers at the Jacksonville Energy Authority power plant in Jacksonville, Florida. These units are capable of burning either 100% coal or 100% petroleum coke or any combination of the two. These units were designed and installed by Foster Wheeler North America Corp.
- Three 262 MWe CFB subcritical boilers at the Turow power plant in Poland. The fuel for these boilers is lignite with moisture content of 45% by weight.
- The largest atmospheric CFB boiler in operation is a 460 MWe unit at a power plant owned by the Polish utility company Południowy Koncern Energetyczny SA (PKE) in Lagisza, Poland.[1] This unit is the world's first supercritical CFB boiler. The primary fuel burned in the unit is Polish bituminous coal. The commercial operation of this unit demonstrates the successful integration of CFB boiler technology with supercritical boiler technology. The unit features include a vertical evaporator with supercritical steam conditions (4,000 psia, 1,050/1,075°F) and a reported overall net plant efficiency of 41.6% (HHV basis).

3.5.1 Pressurized Fluidized-Bed Combustion Systems

Pressurized fluidized-bed combustion (PFBC) systems are FBC systems that operate at elevated pressures (typically pressures of 1–1.5 MPa) and produce a high-pressure gas stream at temperatures that can drive a turbine. Two types of configurations are bubbling beds and a circulating configuration. All operating units in service today are bubbling beds.

In a PFBC, the combustor and hot gas cyclones are all enclosed in a pressure vessel. Both coal and sorbent (for SO_2 emissions reductions) are fed across the pressure boundary, and similar provision for ash removal is provided. For hard coal (i.e., bituminous coal) applications, the coal and limestone can be crushed together, and then fed as a paste, with 25% water. The combustion temperature is between 1,500°F and 1,650°F (800°C to 900°C) has the advantage of less NOx formation than in PC combustion.

[1] Foster Wheeler North America Corp. "Utility CFB Goes Supercritical—Foster Wheeler's Lagisza 460 MWe Operating Experience and New 600–800 MWe Designs." Prepared by James Utt, Arto Hotta, and Stephen Goidich, Foster Wheeler North America Corp., Clinton, NJ, for presentation at Coal-Gen 2009, Charlotte, NC, August 19–21, 2009. Available at http://www.fwc.com/publications/tech_papers/files/TP_CFB_09_12.pdf.

First-generation PFBC designs are based on directly burning crushed coal in the combustor. The high-pressure gas is first expanded through a turbine, and then heat is recovered from the turbine exhaust in a heat recovery steam generator (HRSG) to produce steam, which is used to drive a conventional steam turbine. Figure 3.3 provides a schematic.

There were several demonstration projects (ranging in size from 60 to 130 MWe) implemented during the 1990s in various countries. Japanese equipment manufacturers and electric power companies have led the commercial development of PFBC technology with the construction of several commercial-scale units. They include

- 360 MWe PFBC unit operated by Kyushu Electric Power Company at the Karita Power Station located near Kitakyushu, Japan. The unit began commercial operation in July 2001. The unit uses a supercritical boiler and has a reported net efficiency based on test results of 41.8% HHV.
- 250 MWe PFBC unit operated by Chugoku Electric Power Co., Inc. at Osaki Power Station located near Hiroshima, Japan. Unit 1 began commercial operation in 2000. The planned construction of a second PFBC unit at the facility was canceled in 2008.

More advanced second-generation PFBC system designs use a pressurized carbonizer to first process the feed coal into fuel gas and char (solid material that remains after light gases and tar have been driven out during the initial stage of combustion). The PFBC burns the char to produce steam and to heat combustion air for the combustion turbine. The fuel gas from the carbonizer burns in a topping combustor linked to a combustion turbine, heating the gases to the rated firing temperature of the combustion turbine. Heat is recovered from the combustion turbine exhaust in an HRSG to produce steam, which is used to drive a conventional steam turbine. These systems are also called advanced circulating pressurized fluidized-bed combustion (APFBC) combined cycle systems.

3.6 Gasification

Gasification is a process that uses heat, pressure, and steam to convert any raw material that contains carbon ("carbonaceous") into synthesis gas, which is a gaseous mixture composed primarily of carbon monoxide (CO) and hydrogen (H_2). Carbon monoxide and hydrogen are colorless, odorless, highly flammable gases that can be used to create electricity, chemicals, pure hydrogen, and liquid transportation fuels. Gasification systems are increasingly being relied on to convert raw material feedstocks, such as coal, into useful chemical products.

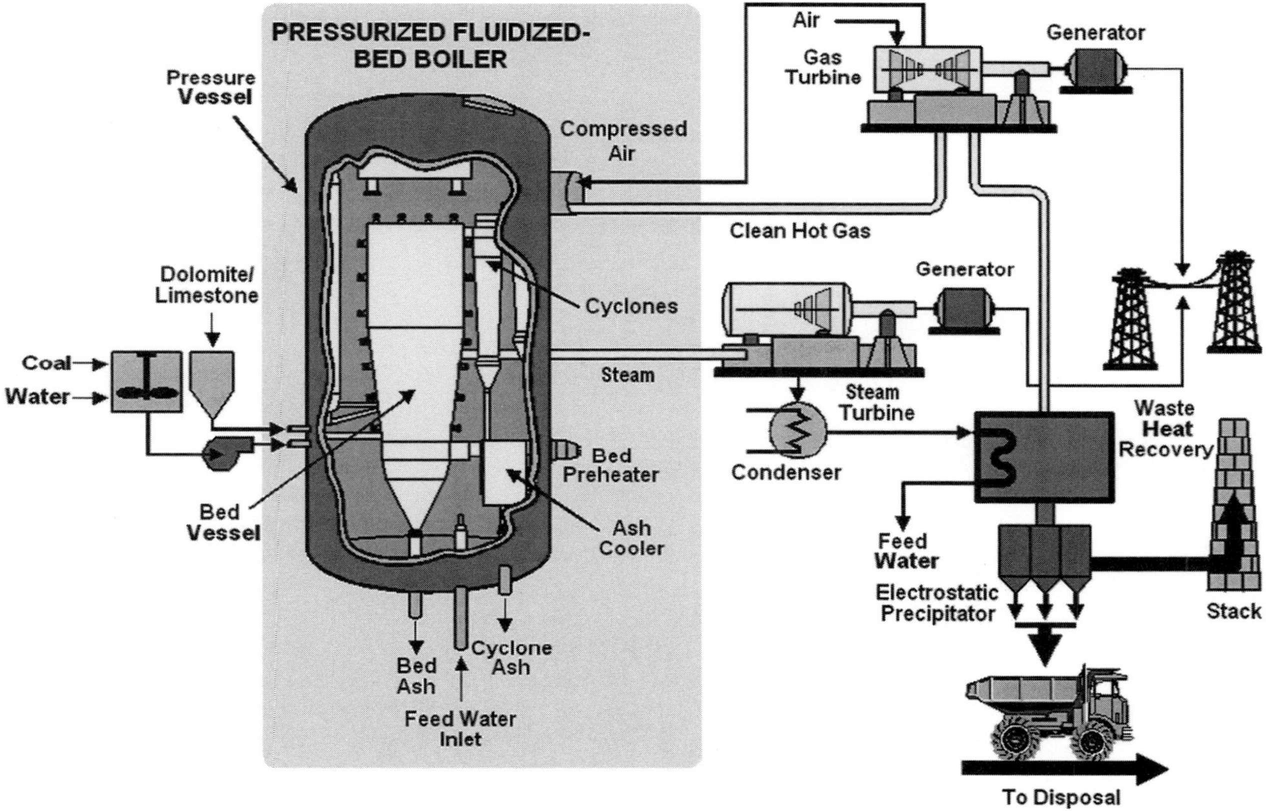

Figure 3.3 Schematic of pressurized fluidized-bed combustion (PFBC) power plant.
Source: *Available and Emerging Technologies for Reducing Greenhouse Gas Emissions from Coal-Fired Electric Generating Units,* U.S. Environmental Protection Agency, Research Triangle Park, North Carolina, October 2010.

Among coal-based power systems, gasification has the smallest environmental footprint. This technology is capable of producing syngas not only from coals having a wide range of heat values but also from low-value carbon feedstocks such as petroleum coke ("Pet coke"), high-sulfur fuel oil, municipal wastes, and biomass. This flexibility increases the economic value of these resources and lowers costs by providing industry with a broader range of feedstock options. It also has a significant advantage of product flexibility. The syngas produced can be converted into different valuable products, ranging from electricity and steam to liquid fuels, basic chemicals, and hydrogen. Integration of multiple products gasification into industrial applications increases commercial opportunities for added revenues since plant operations can focus on the most lucrative products. This provides economies of scale associated with production of multiple commodities and increases opportunities for added revenues to maximize business margins.

Gasification systems result in low emissions of sulfur dioxide (SO_2), particulate matter, and some toxic compounds other than coal contaminants such as mercury, arsenic, selenium, cadmium, and others. This technology does offer a means of capturing and storing or sequestering carbon dioxide (CO_2), a greenhouse gas. The carbon dioxide produced during gasification is present at much higher concentrations and at higher pressures than in streams produced from conventional combustion processes. This factor makes it more economical to capture the CO_2. Theoretically, one can convert the synthesis gas into pure hydrogen using the water–gas shift reaction and then use the hydrogen as an ultra-clean fuel with an exhaust gas water.

Gasification also offers the flexibility of being integrated with other technologies for advanced power generation, such as combustion turbines and possibly solid oxide fuel cells. Envisioned systems are highly efficient because they potentially can capture greater value from each pound of feedstock. Systems using advances in gasification and related components are capable of achieving efficiencies of up to 60%, compared with an efficiency limit of 40% for conventional plants.

The technology itself is by no means new. Gasification was originally developed in the early 1800s for the purpose of producing "town gas" (syngas) from coal for lighting and cooking. Town gas was used to light city streets and homes before natural gas and electricity were introduced for that purpose. By the early 1900s, syngas had fallen out of general public use, replaced by electricity and natural gas. Since the 1920s, gasification has been used mostly for the production of synthetic fuels and chemicals, especially during WWII when petroleum-based fuels (gasoline and oil) were at a premium or unavailable. The Union of South Africa has used gasification to produce a gasoline substitute for transportation fuels because domestic petroleum sources were lacking and oil difficult to import.

In gasification, the feedstock is prepared and fed to a gasifier in either a dry form or as slurry (i.e., mixed with water). The gasifier itself is a reactor vessel where the feedstock reacts with steam and air or oxygen (O_2) at high temperature and pressure in a reducing (oxygen-starved) atmosphere. These conditions produce the syngas, which is made up primarily of carbon monoxide and hydrogen, and smaller quantities of carbon dioxide and methane (CH_4).

The high temperature in the gasifier fuses the remaining inorganic materials into a glassy material, referred to as slag. The slag has the consistency of coarse sand. It is chemically inert and may have a variety of uses in the construction and building industries.

There are different types of gasifiers. Depending on its type, a gasifier may operate at temperatures up to 2,600°F, which is more than 12 times the boiling point of water, and pressures up to 1,000 pounds per square inch, gauge (psig), which is about 68 times that of standard atmospheric pressure. When the feedstock particles enter the gasifier and move through its zones, they undergo three thermal and chemical processes. The first two of these processes are rapid and in succession, followed by a third, slower process. These three processes are known as

- Pyrolysis
- Oxidation
- Reduction

Pyrolysis is a chemical breakdown of complex compounds caused by heat. It occurs as the particles heat up. Volatile substances such as tar, hydrogen, and methane are released, and char, a combustible residue resembling charcoal, is produced, resulting in up to a 70% weight loss for the coal. The properties of the feedstock determine the structure and composition of the char.

Oxidation (best referred to as controlled burning) then occurs as the volatile products and some of the char react with air or oxygen to form carbon dioxide and carbon monoxide. This type of reaction is exothermic (i.e., heat generating).

After pyrolysis and oxidation, reduction then occurs as the char reacts with the carbon dioxide and steam to produce carbon monoxide and hydrogen, with some methane, which together make up the gaseous mixture known as syngas. This type of reaction is endothermic (i.e., consumes the heat). Slag is left as a waste product.

The net result of all the chemical reactions inside a gasifier is that gasification is exothermic—it gives off heat. However, the main product of gasification is the syngas.

Once the raw syngas exits the gasifier, treatment facilities for gas cleaning, gas conditioning, and separation to refine the syngas stream are employed. The following products and byproducts are possible, depending on the desired applications:

- Hot syngas may be used to drive a gas turbine to make electricity.
- A water–gas shift reaction may be used to change the carbon monoxide and water in syngas into hydrogen and carbon dioxide, and so upgrade the heat content of the fuel gas in order to increase the amount of hydrogen that can be separated and used in pure form for a variety of purposes, such as carbon dioxide–free power production, chemical production, and fuel for fuel cells. Water is added to the syngas prior to the water–gas shift reaction to increase the production of hydrogen.
- Syngas may be converted using the Fischer–Tropsch process, which is a catalyzed chemical reaction, into chemicals and liquid hydrocarbon fuels (i.e., fuels containing a combination of hydrogen and carbon), such as lubricating oil and diesel oil, which are easier and safer to transport and pump than flammable gas.
- Chemical products may be manufactured including plastics and urea- or ammonia-based fertilizers.

Gasification can be integrated with modern combustion and steam turbine generators to generate electricity. Natural gas combined cycle systems exceed gasification systems in power-generating efficiency; however, the relatively high price and declining availability of domestic natural gas resources make gasification increasingly attractive as a power-generation option for power producers. This may be an arguable point in the United States because of the potential that fracking technology now offers for recovery of natural gas.

In the power-production scheme, purified syngas from the gasification side of the plant, or "gasification island," is supplied to the power side of the plant, or "power island." The syngas is mixed with air and ignited to drive a combustion turbine directly in a combined thermal and mechanical process that is converted to electrical power by the generator. The electrical power from the generator then is fed to the electrical grid for distribution.

Waste heat from the combustion turbine is used to boil water in the heat recovery steam generator and create the steam to drive a steam turbine with its own generator set. Combined cycle operation makes possible an energy conversion efficiency that is about one-third greater than that possible using a gas turbine only.

This brings us to the subject of the integrated gasification combined cycle (IGCC) power plant. An IGCC power plant uses a coal gasification system to convert coal into a synthetic gas, which is then used as fuel in a combined cycle electric generation process. Gasification processes have been developed using a variety of designs including moving bed, fluidized bed, entrained flow, and transport gasifiers. Figure 3.4 shows a schematic of an IGCC power plant.

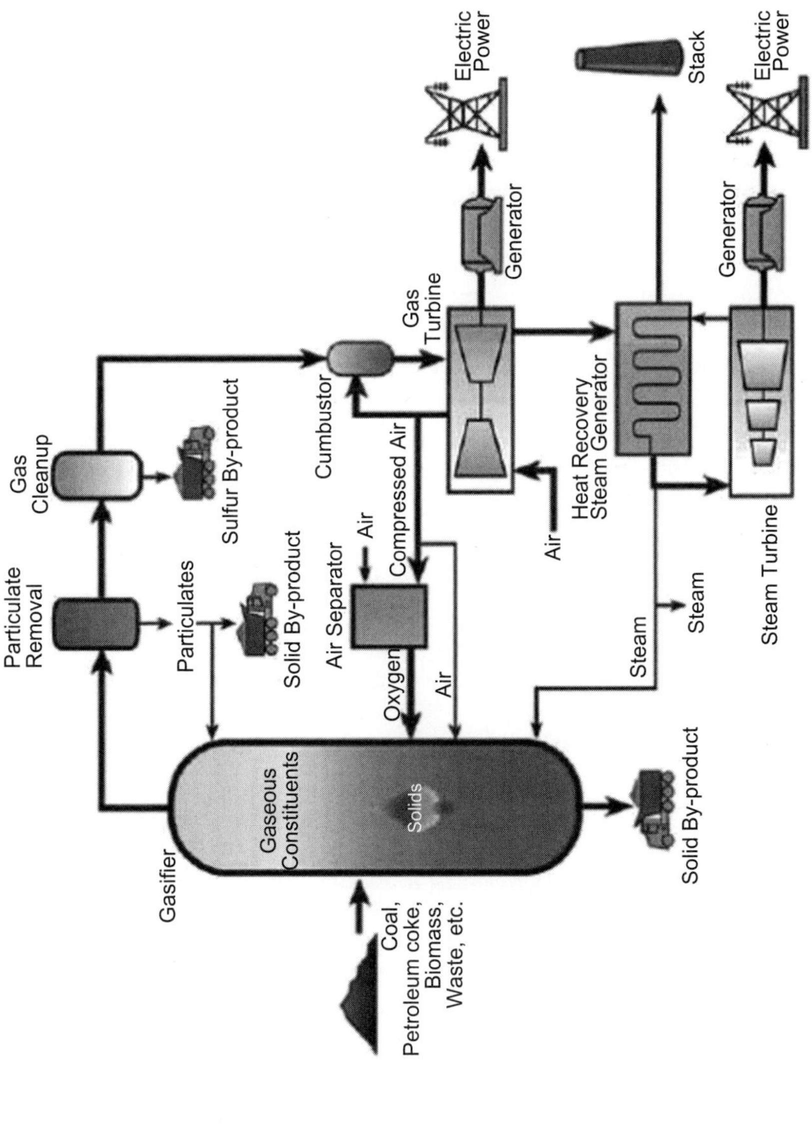

Figure 3.4 Schematic of an integrated gasification combined cycle (IGCC) power plant.
Source: Available and Emerging Technologies for Reducing Greenhouse Gas Emissions from Coal-Fired Electric Generating Units, U.S. Environmental Protection Agency, Research Triangle Park, North Carolina, October 2010.

The hot syngas can be processed to remove sulfur compounds, mercury, and particulate matter (PM) before it is used to fuel a combustion turbine generator to produce electricity. The heat in the exhaust gases from the combustion turbine is recovered to generate additional steam. This steam, along with the steam produced by the gasification process, then drives a steam turbine generator to produce additional electricity.

The efficiency of an IGCC power plant is comparable to the latest advanced PC-fired and CFB EGU designs using supercritical boilers. The advantages of using IGCC technology include greater fuel flexibility (e.g., capability to use a wider variety of coal ranks); potential improved control of PM, SO_2 emissions, and other air pollutants, with the need for fewer post-combustion pollution control devices (e.g., almost all the sulfur and ash in the coal can be removed once the fuel is gasified and prior to combustion); generation of less solid waste requiring disposal; and reduced water consumption when compared to an electric-generating unit using a supercritical boiler.

Like all technologies, IGCC has certain disadvantages, which include additional plant complexity, higher construction costs, and poorer performance at high-altitude locations when compared to a conventional plant using a supercritical boiler. However, IGCC power plants offer the potential for lower control costs of CO_2 emissions because the CO_2 in the syngas can be removed prior to combustion.

Currently operating IGCC plants include the following:

- 253 MWe IGCC plant at the NUON Willem-Alexander Power Plant in Buggenum, Netherlands. The unit began operation in 1994.
- 262 MWe IGCC plant at the Duke Energy Wabash River Power Station in Indiana. The unit began operation in 1995.
- 250 MWe IGCC plant at the Tampa Electric Company (TECO) Polk Power Station in Florida. The unit began operation in 1996.
- 400 MWe IGCC plant at the SUV power plant in Vresova, Czech Republic. The unit began operation in 1996.
- 283 MWe IGCC plant at the ELCOGAS power plant in Puertollano, Spain. The unit began operation in 1998.
- 250 MWe IGCC plant at the Joban Joint Electric Power Company Nakoso Power Station in Iwaki City, Japan. The unit began operation in 2007.

There are two kinds of electric power plants: those designed for "baseload" (i.e., 24-7) operations and those designed to meet demand just at "peak" times. Coal power plants typically provide baseload capacity. Natural gas combined cycle plants generally provide peak capacity. IGCC power plants are being designed primarily for the baseload power market.

Because IGCC will compete most directly against other clean coal technologies, especially the supercritical pulverized coal (SCPC) design, it is useful to compare these two technologies with respect to (1) capital costs and cost of energy produced; (2) environmental impacts; and (3) potential for cost-effective carbon dioxide capture. It is also useful to consider natural gas combined cycle as an additional benchmark.

According to published information developed by GE Energy and Bechtel as part of their "Reference Plant Offering," the total installed cost of a new state-of-the-art SCPC power plant varies with plant size and with the degree of reuse of existing site infrastructure. In a paper presented at the 2005 Gasification Technologies Conference, they stated

> *At a nominal plant size of 600 MWe, the price for a SCPC plant in the Ohio River Valley would typically fall within the 1,200 to 1,460 $/kW range. Presently, IGCC plants cost 20 to 25 percent more than a comparable SCPC power plant at any given site. However, the GE Energy-Bechtel alliance's next-generation IGCC plants are targeted to enter the marketplace with a price that will reduce that capital cost premium to the range of 10 percent.*[2]

The environmental benefits of IGCC partially close the cost-of-electricity gap caused by the 20% to 25% capital cost premium. Assuming a 10% capital cost premium and the inherent improved environmental performance of IGCC plants versus SCPC plants, results are near parity for the cost of electricity for next-generation IGCC and SCPC projects.

Criteria Air Pollutants. Air pollutants emitted from vehicles and stationary sources such as power plants are regulated in the United States and other nations. These so-called criteria air pollutants include sulfur dioxide (SO_2), nitrogen oxides (NOx), particulate matter (PM), and carbon monoxide (CO). The U.S. Environmental Protection Agency has established New Source Performance Standards for large-scale, combustion-based power plants that use coal. As illustrated in the following paragraphs, present and next-generation IGCC facilities (1) meet or exceed the EPA's standards; (2) emit fewer criteria air pollutants than SCPC plants; and (3) compare favorably with natural gas combined cycle (NGCC) facilities.

In terms of environmental footprints, we may start with mercury because there has been a long-running debate in the United States about the regulation of mercury from power plants. The EPA issued a final Clean Air Mercury rule in 2005 that instituted a two-phase reduction in mercury emissions. The first phase was proposed by the EPA on May 3, 2011 and a final rule was made effective April 16, 2012. This final rule will reduce emissions of HAP, including mercury, from the electric power industry. The new technology required to reduce HAP will also reduce $PM_{2.5}$ and SO_2 emissions as

[2] http://www.clean-energy.us/facts/igcc.htm#projects

a cobenefit. The second phase (due by 2018) will see mercury emissions limited further through a national limit (15 tons per year), pro rata limits, and a "cap and trade" system. The December 2002 Department of Energy (DOE) study titled "Major Environmental Aspects of Gasification-Based Power Generation Technologies" reached the following conclusions concerning mercury in IGCC and other clean-coal power systems:

> *Compared with combustion-based power plants, IGCC plants have a major advantage when it comes to mercury control. Commercial methods have been employed for many years that remove trace amounts of mercury from natural gas and gasifier syngas. Both molecular sieve technology and activated carbon beds have been used for this purpose, with 90 to 95 percent removal efficiency reported.... A recent DOE cost study was conducted for applying a packed-bed carbon adsorption system to an IGCC plant. Based on an 18-month carbon replacement cycle and 90 percent reduction of mercury emissions, the total cost of mercury reduction is estimated to be $3,412 per pound of mercury removed, which is projected to be about one-tenth the cost of flue gas-based [SCPC] mercury control.*[3]

In terms of volumes of solid waste material produced, as well as the potential for leaching of toxic substances into soil and groundwater, IGCC has demonstrated reduced environmental impact compared with similarly sized coal combustion-based power plants. The largest solid waste stream produced in an IGCC facility is slag (or bottom ash in some designs). As described previously, slag is a black, glassy, sand-like material that can be a marketable byproduct, and leachability data obtained from different gasifiers shows that gasifier slag is highly "nonleachable."

This brings us to the greenhouse gas, carbon dioxide. Of particular concern is the large volume of carbon dioxide (CO_2) emitted by coal-based power plants. CO_2 traps heat in the atmosphere, and many scientists believe that increased greenhouse gas emissions will soon lead to global climate change. While it is theoretically possible to capture CO_2 from conventional coal-fired plants, to do so will be prohibitively expensive. The cost of CO_2 capture and sequestration in an IGCC plant is also projected to be expensive, but it is a much smaller percentage of the cost of producing electricity. The principal reason for this cost differential is that the CO_2 in an IGCC plant is separated from the syngas before combustion, whereas the CO_2 in a conventional coal plant is removed from the post-combustion exhaust gases.

[3] Major Environmental Aspects of Gasification-Based Power Generation Technologies, Dec 2002. DOE/NETL. p. ES-5 <http://www.netl.doe.gov/technologies/coalpower/gasification/pubs/pdf/final%20env.pdf>

3.7 Comparing Energy Costs

Since the 1980s, the amount of electricity produced from coal globally has more than tripled. This is mainly due to the relatively low cost of building and running coal-fired power plants. Climate change policies and environmental regulations, however, have placed pressures on the economics of electricity from coal.

The overall cost of generating electricity is one of the most important factors determining the choice of technology for new power generation. This is true in both developing countries where universal access to energy services is still a challenge, as well as in the established economies where electricity prices are an important variable for the standard of living.

Numerous studies by the European Commission, and the U.S. Congressional Budget Office show that coal power plants provide electricity at a lower cost than do nuclear or gas plants. This is also confirmed by levelized generation cost studies, such as the one carried out regularly by the International Energy Agency (IEA), which considers all the costs over the power plant lifetime. According to IEA statistics, coal-based electricity is, on average, 7% less costly than gas and around 19% more economical than nuclear.

IEA and European Commission studies also show that onshore wind costs between US$50 and US$156 per MWh and solar photovoltaics (PVs) between US$226 and US$2,031. In certain locations, hydro resources can produce electricity at a cost comparable to that for coal; however, estimates vary greatly according to geographic conditions, and the final price can be as high as US$240 and US$262 per MWh. In comparison, electricity from coal costs between US$56 to US$82 per MWh. The economies of scale simply favor coal.

In the IEA publication on levelized costs of electricity, regional costs of nuclear vary between US$30 per MWh in China and US$60 in Europe. This is much lower than the average cost for nuclear quoted by other international studies, such as those by the European Commission. These variations come from differences in accounting for costs related to nuclear waste disposal and to the decommissioning of aging power plants.

Cost estimates and relative competitiveness of other energy resources can also vary depending on whether the associated infrastructure spending needs are taken on board. As an example, according to IEA statistics, capital intensity of investment in the entire natural gas chain is six times higher than for coal. This also means that there is a cost associated with fuel switching that is not transparent in simple, power-plant-level cost estimates. In the case of the United States, an additional US$348 billion investment in pipeline capacity and further US$12.5 billion investment in storage capacity would be required if the United States were to switch from coal to natural gas. These cost sensibilities need to be taken into account by policymakers if informed choices are to be made about the future energy mix.

The pricing of CO_2 emissions is expected to affect coal more than gas or nuclear power. However, with the 2010 failure of the U.S. cap-and trade bill, carbon pricing policies seem to have significantly lost momentum. There are, however, other variables, such as commodity prices and interest rates, which, like carbon pricing, can affect the competitiveness of energy resources.

According to a report on the future role of coal in Europe by Prognos,[4] a Swiss-based consultancy, coal can compete with gas at carbon prices of up to US$30, and the competitiveness of coal power goes up as the price of natural gas increases. As of this writing, natural gas prices are historically low. Recent increases in shale gas extraction in the United States have led to a global oversupply of natural gas and a drop of gas prices. However, there is no certainty as to how long the situation will continue. On the other hand, if both coal and gas prices were to double, coal would be over 35% less expensive than gas for electricity production, even with a carbon price of US$30 per metric ton of CO_2.

Nuclear power is more competitive than coal and gas in a number of countries; however, IEA statistics show that the cost of nuclear power varies greatly with changes in interest rates. A hike from 5% to 10% produces an increase of 50% in the cost of nuclear power. In comparison to this, a similar increase in interest rates for coal would result in 25% of additional costs. Also, due to the long lead times and high capital intensity, investments in nuclear power are generally considered a higher risk and, as a result, higher interest rates typically apply to investments in nuclear power. This is not reflected in some of the international cost assessment studies that apply similar interest rates to all energy technologies.

Early cost estimates for carbon capture and storage (CCS) indicate that electricity can be produced from abated coal at a cost that competes with other low-carbon technologies. The first CCS demonstration projects generate electricity at half the price of solar PVs. An independent report for the Global CCS Institute estimates that the current cost of producing electricity from abated coal is US$127 per MWh. This is comparable to the average price of US$125 paid by the European governments for wind energy through the feed-in tariffs.

Abated coal has the potential to become more competitive as the IEA expects the additional cost of CCS to go down from US$40 to US$30 per MWh before 2030. CO_2 could also follow the same path as other coal combustion products and become a commercial product with a market value. The ongoing economic

[4] http://www.euracoal.be/componenten/download.php?filedata=1208519374.pdf&filename=prognos_FutureCoal_070822_final_kurz.pdf&mimetype=application/pdf

stagnation in many developed countries and the enduring challenge of electrification in the developing world make it necessary to consider clean but also affordable and reliable electricity generation. IEA analysis implies that CCS technology can reduce the overall cost of mitigating climate change by 70%. Coal power plants are expected to account for 65% of CCS deployment by 2050. This makes coal one of the few energy sources that can address both the challenge of climate change and development.

4 Carbon Capture and Sequestration

Chapter Outline

4.1 Introduction

Coal is composed of carbon, hydrogen, oxygen, nitrogen, sulfur, moisture, and incombustible mineral matter (i.e., ash). When coal is burned, the carbon and hydrogen oxidize to form the primary combustion products of CO_2 and water. Other combustion products such as NOx, SO_2, CO, and particulate matter (PM) are formed in varying amounts.

The principal greenhouse gases (GHGs) that enter the atmosphere because of human activities are CO_2, nitrous oxide (N_2O), methane (CH_4), hydrofluorocarbons (HFCs), perfluorocarbons (PFCs), and sulfur hexafluoride (SF_6). Of these, CO_2 is by far the most abundant GHG emitted from power production by coal utilization.

To optimize overall efficiency for a given coal-fired power plant, the unit is operated under conditions such that nearly all the fuel carbon is converted to CO_2 during the combustion process. Methane is emitted during the mining and transport of coal but is not a significant byproduct of electric-generating coal combustion. Fluorinated gases are not formed by coal combustion.

The formation of N_2O during the combustion process results from a complex series of reactions, and its formation depends on many factors. However, the formation of N_2O is minimized when combustion temperatures are kept high and excess air is kept to a minimum.

PC-fired boilers are typically operated at conditions such that they are not significant sources of N_2O emissions. However, fluidized-bed combustion

Handbook of Pollution Prevention and Cleaner Production Vol. 4. DOI: http://dx.doi.org/10.1016/B978-1-4377-7815-1.00004-7

(FBC) boilers can have measurable N_2O emissions, resulting from the lower combustion temperatures and the use of selective noncatalytic reduction (SNCR) to reduce NOx emissions. Operating factors impacting N_2O formation include combustion temperature, excess air, and sorbent feed rate. The N_2O formation resulting from SNCR depends on the reagent used, the amount of reagent injected, and the injection temperature.

There are over 50,000 power plants throughout the world. These power plants constitute the single largest emitting industry for CO_2 emissions. With the increased emphasis on greenhouse gas abatement, it is critical that these power plants undertake efforts to reduce their CO_2 emissions, or capture the CO_2 emitted and sequester it. This chapter provides an overview of GHG control schemes and technologies.

4.2 Factors Influencing GHG Emissions

The level of CO_2 emissions that can be released from a given coal-fired electric-generating plant depends on the type of coal burned, the overall efficiency of the power-generation process, and use and condition of air pollution control devices employed. Table 4.1 provides a summary of the factors that have the greatest impact. Because of these factors, coal-fired units that are identical in design but operated by different utility companies in different locations may have different efficiencies. The level of effectiveness of a given GHG control technology used to improve the efficiency at one coal-fired facility may not necessarily directly transfer to a coal-fired facility at a different location.

4.3 Emissions Trading Scheme

The European Union (EU) Emissions Trading Scheme (ETS) is the largest greenhouse gas emissions trading scheme of its kind in the world. The scheme requires the annual reporting of CO_2 emissions by mass, similar to the routine measurement and reporting undertaken in countries such as the United States. Specific emissions reporting per unit of production (e.g., gCO_2/kWh) is not required. Emission calculations are based on fuel used and agreed oxidation factors. Reporting of figures on the basis of actual oxidation factors, rather than the assumed values, is permitted where actual values are available and can be verified.

Participating countries have a national allocation, agreed by the European Commission (EC), that is allocated or auctioned to major emitters by

Table 4.1 Factors Influencing GHG Emissions

Factor	Description
Coal Rank	The amount of heat released by coal combustion depends on the amounts of carbon, hydrogen, and oxygen present in the coal and, to a lesser extent, on the sulfur content. The ratio of carbon to heat content depends on these heat-producing components of coal, and these components vary by coal rank. In general, anthracite emits the largest amount of CO_2 per million Btu (MMBtu), followed by lignite, subbituminous coal, and bituminous coal. However, for a given coal rank, there is variation in the CO_2 emission factor depending on the coal bed from which the coal is mined.
Moisture Content	In addition to the lower CO_2 emissions rate per unit of heat input (lbs CO_2/MMBtu), due to the inherent moisture in subbituminous and lignite coals, all else being equal, a bituminous coal-fired boiler is more efficient than a corresponding boiler burning subbituminous or lignite coal. Switching from a low- to a high-rank coal will tend to lower GHG emissions from the utility stack. However, overall GHG emissions might not be lowered by switching to bituminous coal. All coal mining operations release coal bed methane to the atmosphere during the mining process. Some bituminous coal reserves release significant amounts of methane, which could, in theory, offset GHG savings. Additional factors when considering overall GHG emissions include the fuel needs to mine, process, and transport the coal.
Thermal Efficiency	As the thermal efficiency of a coal-fired unit is increased, less coal is burned per kilowatt-hour (kWh) generated, and there is a corresponding decrease in CO_2 and other air emissions. There is no standardized procedure for continuous online measurement of coal-fired thermal efficiency. A near approximation performed under the EPA's Acid Rain Program collects heat input and gross megawatt output on an hourly basis to calculate gross heat rate. The heat input is derived from standardized continuous emission monitors, while the utility supplies gross megawatt output. The electric energy output as a fraction of the fuel energy input expressed in percentage is a commonly used practice for reporting the efficiency of a coal-fired unit. The greater the output of electric energy for a given amount of fuel energy input, the higher the efficiency for the electric-generation process. Heat rate is another common way to express efficiency. Heat rate is expressed as the number of British thermal units (Btu) or kilojoules (kJ) required to generate a kilowatt-hour of electricity. Lower heat rates are associated with more efficient power-generating plants. Although the same basic formula is used to calculate efficiency for coal-fired units, there are different methodologies for measuring the appropriate parameters. For example, the varying accuracy of the different methodologies can cause discrepancies in measuring the heating value of the coal burned. Although there is a direct inverse correlation between coal-fired electric-generating unit (EGU) efficiency and CO_2 emissions, other factors must be considered when comparing the effectiveness of GHG control technologies to improve the efficiency of a given coal-fired unit. The actual overall efficiency that a given coal-fired unit achieves is determined by the interaction of a combination of site-specific factors that impact efficiency to varying degrees. These factors include

(*Continued*)

Table 4.1 Factors Influencing GHG Emissions—*cont'd*

Factor	Description
	thermodynamic cycle, coal rank and quality, plant size, pollution control systems (the electric power consumed by air pollution control equipment reduces the overall efficiency of the plant), operating and maintenance practices, and the cooling system (the temperature of the cooling water entering the condenser can have impacts on steam turbine performance; once-through cooling systems can have an efficiency advantage over recirculating cooling systems [e.g., cooling towers]. However, once-though cooling systems typically have larger water-related ecological concerns than recirculating cooling systems. Other factors influencing thermal efficiency include the plant's geographic location (elevation and seasonal ambient temperatures at the facility site potentially may have a measureable impact on efficiency). At higher elevations, air pressure is lower and less oxygen is available for combustion per unit volume of ambient air than at lower elevations. Cooler ambient temperatures can increase the overall efficiency by increasing the draft pressure of the boiler flue gases and the condenser vacuum and by increasing the efficiency of a condenser recirculating cooling system); Load generation flexibility requirements (operating as a baseload unit is more efficient than operating as a load cycling unit to respond to fluctuations in customer electricity demand); Equipment manufacturers (e.g., the efficiency specifications of major components such as boilers, turbines, and electrical generators provided by equipment manufacturers can affect overall efficiency); Plant components (e.g., using the optimum number of feedwater heaters, high-efficiency electric motors, variable-speed drives, better materials for heat exchangers, etc., tend to be more efficient).
SO_2 Controls	SO_2 emissions from new coal-fired units, or retrofitting of an existing facility without specific SO_2 controls, are controlled using flue-gas desulfurization (FGD) technology to remove the SO_2 before it is vented to the atmosphere. The selection of the type of FGD technology will impact overall GHG emissions. All FGD processes require varying amounts of electric power to operate, which contributes to the overall parasitic load of the unit. The FGD parasitic load requirements are between 1% and 2% of the gross output of the facility. In addition, some FGD processes use carbon-containing reagents (e.g., carbonates) that form CO_2 as a byproduct of the chemical reactions of the reagent with SO_2. For a typical unit, the CO_2 that is chemically created in a scrubber adds an additional 1% to the overall GHG emissions, but it can be as high as 3% for facilities burning high-sulfur coals. However, from an overall GHG emissions standpoint, the use of FGD technologies that do not form byproduct CO_2, such as lime-based scrubbers, does not necessarily reduce emissions. Lime is manufactured by heating limestone in the absence of oxygen to remove a molecule of CO_2 ($CaCO_3$ + heat $\rightarrow$ $CaO + CO_2$). Unless the CO_2 is sequestered at the lime production facility, overall GHG emissions will be similar.

Source: *Available and Emerging Technologies for Reducing Greenhouse Gas Emissions from Coal-Fired Electric Generating Units*, U.S. Environmental Protection Agency, Research Triangle Park, North Carolina, October 2010.

governments. Emitters must either comply with their allocated CO_2 cap using the allowances they hold, buy additional allowances to cover their requirements, or pay a severe fine for exceeding their allocation (and buy the missing allowances). Allowances are traded and surplus allowances can be sold. Since trading began in 2005, market prices have fluctuated significantly.

Following an amendment to the ETS Directive, agreed on December 17, 2008, a greater share of allowances must be auctioned from 2013, with full auctioning generally required in the electricity sector. The scheme does not provide a direct means of comparing CO_2 emission rates of different technologies.

Article 14 of the ETS Directive requires the European Commission to elaborate guidelines for monitoring and reporting greenhouse gas emissions under the ETS. Monitoring and reporting guidelines were adopted in January 2004 with Commission Decision 2004/156/EC.14 Article 14, which also requires member states to ensure that emissions are monitored in accordance with the legally binding guidelines. Since 2004, the guidelines have been subject to a number of revisions (Commission Decisions 2007/589/EC, 2009/73/EC, 2009/339/EC and a draft amendment covering carbon capture and storage [CCS]).

Most reporting systems in use around the world report on the basis of CO_2 emissions and use factors to report other greenhouse gas emissions as their CO_2 equivalent (CO_2e). Sometimes CO_2 emissions are quoted in terms of carbon (C), or confusingly are stated as "carbon" when they are in fact CO_2. The equivalence in mass terms between carbon dioxide and carbon is simply the ratio of their molecular masses CO_2:C, this being 3.6632.

Input-based emission calculations and limits for the mass of CO_2 emitted per unit of input energy, expressed in units such as tCO_2/GJ or lb/MMBtu coal, create a poor comparison of specific emission rates. They imply that producing the same total emissions from the same quantity of fuel represents equivalent performance. In reality, although a more efficient plant consuming the same mass of fuel as a less efficient plant creates the same total quantity of CO_2, it does so with the benefit of producing more useful output energy. Emission standards based on useful energy output (e.g., tCO_2/GWh) are therefore important, since they capture the benefits of higher efficiency and incentivize the development and implementation of cleaner, more efficient technologies.

4.4 Carbon Capture and Storage (CCS) Technologies

The terms "carbon capture and storage," and "carbon capture and sequestration" refer to technology aimed at releasing CO_2 into the atmosphere from fossil fuel use. The process is based on capturing carbon dioxide (CO_2) from large point sources (i.e., stacks). It encompasses CO_2 scrubbing of from ambient air. CO_2 has been injected into geological formations for various purposes, but in general the long-term storage of CO_2 is a relatively new concept.

Capturing and compressing CO_2 may increase the fuel needs of a coal-fired CCS plant by 25%–40%. These and other system costs are estimated to increase the cost of the energy produced by 21%–91% for purpose-built plants. Applying the technology to existing plants would be even more expensive, especially if they are far from a sequestration site.

Storage of the CO_2 is envisaged either in deep geological formations, in deep ocean masses, or in the form of mineral carbonates. Deep ocean storage risks greatly increasing the problem of ocean acidification, which is a problem that also originates from the excess of carbon dioxide already in the atmosphere. Geological formations are currently considered the most promising sequestration sites.

The National Energy Technology Laboratory (NETL) has noted that North America has sufficient storage capacity for more than 900 years' worth of carbon dioxide at current production rates.[1] It is important to note that long-term predictions about underground storage security are difficult and have high uncertainty; plus there exists the probability that CO_2 might leak from the storage into the atmosphere.

The most practical current method for CO_2 capture is scrubbing. There are essentially three different types of technologies for scrubbing: post-combustion, pre-combustion, and oxyfuel combustion. See Table 4.2 for an overview.

Post-combustion capture involves the removal of CO_2 after combustion of the fossil fuel. Carbon dioxide is captured from flue gases using a scrubbing liquid.

Pre-combustion is an alternate method that is widely applied in fertilizer, chemical, gaseous fuel (H_2, CH_4), and power production. In these cases, the fossil fuel is partially oxidized, for instance, in a gasifier. The syngas (CO and H_2O) is shifted into CO_2 and more H_2. The resulting CO_2 can be captured from a relatively pure exhaust stream. The H_2 can be used as fuel; the carbon dioxide is removed before combustion takes place. The CO_2 is removed after combustion of fossil fuels but before the flue gas is expanded to atmospheric pressure. This scheme is applied to new fossil-fuel-burning power plants or to existing plants where repowering is an option.

Oxyfuel combustion involves the combustion of the fuel in oxygen instead of air. To limit the resulting flame temperatures to levels common during

[1] NETL, *Carbon Sequestration Atlas,* 2007.

conventional combustion, cooled flue gas is recirculated and injected into the combustion chamber. The flue gas is composed largely of carbon dioxide and water vapor. The water vapor can be condensed through cooling. This produces an almost pure carbon dioxide stream that can be transported to the sequestration site and stored. Power plant processes based on oxyfuel combustion are referred to as "zero emission" cycles because the CO_2 stored is not a fraction removed from the flue-gas stream (as in the cases of pre- and post-combustion capture) but the flue-gas stream itself. A certain fraction of the CO_2 generated during combustion will inevitably end up in the condensed water. To warrant the label "zero emission," the water would thus have to be treated or disposed of appropriately. The technique is promising, but the initial air separation step requires substantial energy, thereby making it inefficient.

Another technology under development is known as chemical looping combustion (CLC). Chemical looping utilizes a metal oxide as a solid oxygen carrier. Metal oxide particles react with a solid, liquid, or gaseous fuel in a fluidized bed combustor. This generates solid metal particles and a mixture of carbon dioxide and water vapor. The water vapor is condensed, leaving pure carbon dioxide, which can then be sequestered. The solid metal particles are circulated to another fluidized bed, where they react with air, producing heat and regenerating metal oxide particles that are recirculated to the fluidized bed combustor. A variant of CLC is calcium looping, which uses the alternating carbonation and then calcination of a calcium-oxide-based carrier as a means of capturing the CO_2.

Capturing CO_2 is only one part of the problem. After capture, the CO_2 needs to be transported to suitable storage sites. This has to be done by pipeline. Roughly 6,000 km of CO_2 pipelines in the United States are used to transport CO_2 to oil production fields, where it is then injected into older fields to extract oil. The injection of CO_2 to produce oil is generally called enhanced oil recovery (EOR). In addition, there are several pilot programs in various stages to test the long-term storage of CO_2 in non-oil-producing geologic formations.

According to Congressional Research,

> *There are important unanswered questions about pipeline network requirements, economic regulation, utility cost recovery, regulatory classification of CO_2 itself, and pipeline safety. Furthermore, because CO_2 pipelines for enhanced oil recovery are already in use today, policy decisions affecting CO_2 pipelines take on an urgency that is unrecognized by many. Federal classification of CO_2 as both a commodity (by the Bureau of Land Management) and as a pollutant (by the Environmental Protection Agency) could potentially create an immediate conflict which*

Table 4.2 Overview of Scrubbing Technologies

Technology	Overview
Pre-combustion	The first of the three capture technologies—pre-combustion—allows operators of power plants to capture CO_2 and maximize power output. With this method, an air separation unit produces a stream of almost pure oxygen. The oxygen flows into the gasifier and reacts with pulverized coal to form synthetic gas, or "syngas." Steam is added to the syngas in a shift reactor, converting the carbon monoxide to hydrogen and CO_2. Through use of a physical wash, the CO_2 is then captured from the gas stream and, after compression and dehydration, is ready for transport and storage. Today, the hydrogen is burned to power turbines and makes electricity; tomorrow it could also be used as fuel for transport. The flue gas that results from the hydrogen-powered turbines passes through a heat recovery steam generator (HRSG), which powers steam turbines, reusing this energy and optimizing energy output.
Post-combustion	Post-combustion has the advantage of being able to be installed on both existing and future power plants—of vital importance given that the average power plant operates for 40 years, sometimes longer. A mixture of coal (or gas) and air is blasted into the boiler and ignited. Many power stations "wash" the coal before it is fed into the boiler. "Washing" actually refers to a process that involves passing coal through a series of liquids of varying densities, removing many of the impurities. The heat from the combustion of the coal and air mixture generates steam, which drives the turbine. The flue gas, a byproduct of burning coal, is removed from the boiler and filtered. The steam powers turbines to generate electricity, which is then transmitted to the distribution grid. Once the steam has passed through the turbine, it arrives at a condenser. This unit uses cool

	water to condense the steam back into water, allowing it to be piped back into the boiler and be reheated. This is the first of several "cleaning" processes that the flue gas will pass through. At this point, small particles called "fly ash" are removed from the gas. Sulfur is also removed from the flue gas before it enters the CO_2 absorber, where it needs to be cooled. This stage, using water, lowers the temperature of the gas. Here, the gas stream is typically passed though a liquid sorbent, which reacts with the CO_2, chemically binds with it, and removes it from the flue gas. Once the CO_2 is captured, the sorbent is moved to a desorber to be "regenerated" which usually involves heating the sorbent to release the captured CO_2.
Oxyfuel	Oxyfuel consists of burning fuel in a mixture of pure oxygen and recirculated flue gas instead of air. This is done to increase the CO_2 concentration in the flue gas, thereby making it more efficient to remove before processing for transport and storage. This method deploys an air separation unit that removes nitrogen from the air, producing oxygen. This is injected—alongside the fuel—into a boiler where combustion takes place. Steam is generated and used to power turbines and make electricity. Meanwhile, the flue gas, CO_2, and water vapor are recirculated to control boiler temperature and gradually cooled. This leaves the captured CO_2 to be compressed and dehydrated, ready for transport and storage. To date, the only two examples of oxyfuel combustion are Swedish utility Vattenfall's 30 MW pilot plant in Schwarze Pumpe, Germany, and energy company Total's Lacq project in southwestern France.

may need to be addressed not only for the sake of future CCS implementation, but also to ensure consistency of future CCS with CO_2 pipeline operations today. [2]

Various forms have been conceived for permanent storage of CO_2. They include gaseous storage in deep geological formations (including saline formations and exhausted gas fields), liquid storage in the ocean, and solid storage by reaction of CO_2 with metal oxides to produce stable carbonates.

Geological storage is referred to as geosequestration. This method involves injecting carbon dioxide, generally in supercritical form, directly into underground geological formations. Oil fields, gas fields, saline formations, unmineable coal seams, and saline-filled basalt formations have been proposed as storage sites. Various physical (e.g., highly impermeable caprock) and geochemical trapping mechanisms are required to prevent the CO_2 from escaping into the atmosphere.

CO_2 is currently injected into declining oil fields to enhance oil recovery. The geology of hydrocarbon reservoirs is generally well understood, and storage costs may be partly offset by the sale of additional oil that is recovered. The disadvantages of using old oil fields are their geographic distribution and their limited capacity, as well as the fact that subsequent burning of the additional oil so recovered will offset much or all of the reduction in CO_2 emissions.

Another option is to store captured CO_2 in unmineable coal seams. Technical feasibility really depends on the permeability of the coal bed, but conceptually it is attractive because the CO_2 would be adsorbed by the coal. In the process of adsorption, the coal releases previously adsorbed methane, and the methane can be then be recovered. This is known as enhanced coal bed methane recovery. The sale of the methane can then be used to offset a portion of the cost of the CO_2 storage. Burning the resultant methane, however, would produce CO_2, which would negate some of the benefit of sequestering the original CO_2.

Another storage option involves the application of saline formations that are composed of concentrated mineralized brines. Saline aquifers have been used for storage of chemical waste in a few instances. The main advantage of saline aquifers is their large potential storage volume and their common occurrence. The major disadvantage of saline aquifers is that relatively little is known about them, especially compared to oil fields. To keep the cost of storage low, the geophysical exploration may be limited, resulting in larger uncertainty about the aquifer structure. Unlike storage in oil fields or coal bed

[2] Adam Vann and Paul W. Parfomak, "CRS Report for Congress: Regulation of Carbon Dioxide (CO_2) Sequestration Pipelines: Jurisdictional Issues," Updated April 15, 2008 (Order Code RL34307).

seams, no side product will offset the storage cost. Leakage of CO_2 back into the atmosphere may be an issue in saline aquifer storage.

Another proposed form of carbon storage is in the oceans. Four conceptual approaches are being considered. They include the following:

1. Dissolution, which involves CO_2 injection by ship or pipeline into the ocean water column at depths of 1,000–3,000 meters, forming an upward-plume, and the CO_2 subsequently dissolves in seawater.
2. Injection through "lake" deposits, by injecting CO_2 directly into the sea at depths greater than 3,000 meters, where high pressure liquefies CO_2, making it denser than water, thus forming a downward-plume that may accumulate on the sea floor as a "lake." This approach is expected to delay dissolution of CO_2 into the ocean and atmosphere, possibly for millennia.
3. The use of a chemical reaction to combine CO_2 with a carbonate mineral (such as limestone) to form bicarbonate(s)—for example, $CO_2 + CaCO_3 + H_2O \rightarrow Ca(HCO_3)_2(aq)$. Note that the aqueous bicarbonate solution must not be allowed to dry out, or else the reaction will reverse.
4. Storage of CO_2 in solid clathrate hydrates already existing on the ocean floor.

Environmental effects of oceanic storage are generally negative and not well understood. Large concentrations of CO_2 would be toxic to ocean organisms. Another problem is that dissolved CO_2 would eventually equilibrate with the atmosphere, so the storage would not be permanent. Also, part of the CO_2 would react with the water to form carbonic acid (H_2CO_3), thereby increasing the acidity of the ocean water. Even though life appears to be sparse in the deep ocean basins, energy and chemical effects in these deep basins could have implications of a global nature. Considerably more research is needed to better define the extent of the potential problems. The bicarbonate approach would reduce the pH effects and enhance the retention of CO_2 in the ocean, but it would also increase the costs and other environmental impacts.

An additional method of long-term ocean-based sequestration is to gather crop residue such as corn stalks or excess hay into large weighted bales of biomass and deposit it in the alluvial fan areas of the deep ocean basin. Dropping these residues in alluvial fans would cause the residues to be quickly buried in silt on the sea floor, sequestering the biomass for very long time spans. Alluvial fans are prevalent in all of the world's oceans and seas where river deltas fall off the edge of the continental shelf. Examples are the Mississippi alluvial fan in the Gulf of Mexico and the Nile alluvial fan in the Mediterranean Sea. On the other hand, biomass and crop residues form an important component of topsoil and sustainable agriculture. Removing them from the terrestrial equation creates other problems. If fertilized crops were used, it would exacerbate nutrient depletion and increase dependence on chemical fertilizers and, therefore, petrochemicals, thus defeating the original intentions of reducing CO_2 in the atmosphere. Less-expensive cellulosic energy crops would be used, and these are typically unfertilized, although it is likely that petrochemicals would still be used for harvesting and transport.

Mineral storage is yet another concept being considered. In this process, CO_2 can be exothermically reacted with metal oxides, which in turn produces stable carbonates. This process occurs naturally, resulting in the formation of surface limestone. This reaction does take a long time, but reaction rates can be enhanced by increasing the kinetics through higher temperatures and/or pressures or by pretreatment of the minerals. The economics of mineral carbonation at scale are now being tested in a world first pilot plant project based in Newcastle, Australia. New techniques for mineral activation and reaction have been developed the GreenMag Group and the University of Newcastle and funded by the New South Wales and Australian governments to be operational by 2013.[3]

4.4.1 Safety and Environmental Setbacks

There are serious concerns over the various proposed CCS options. A major one is whether leakage of stored CO_2 will compromise CCS as a climate change mitigation option. For well-selected, -designed, and -managed geological storage sites, the risks are likely comparable to those associated with current hydrocarbon activity. However, there are far too many examples of poor industry practices that have placed the public at risk. The Berkel en Rodenrijs, Netherlands, incident in December 2008 was an example, where a modest release of CO_2 from a pipeline under a bridge resulted in the deaths of some ducks sheltering there.[4] To measure accidental CO_2 releases more accurately and reduce the risk of fatalities through this type of leakage, the implementation of CO_2 alert meters would be needed.

In 1986 a large leakage of naturally sequestered CO_2 rose from Lake Nyos in Cameroon and asphyxiated 1,700 people. The Lake Nyos disaster resulted from a volcanic event, which suddenly released as much as a cubic kilometer of CO_2 gas from a pool of naturally occurring CO_2 under the lake in a deep narrow valley.

For ocean storage, the retention of CO_2 would depend on the depth. Reported estimates of 30%–85% of the sequestered carbon dioxide would be retained after 500 years for depths 1,000–3,000 meters. Mineral storage is not regarded as having any risks of leakage.

[3] See www.dpi.nsw.gov.au. June 6, 2010. http://www.dpi.nsw.gov.au/minerals/resources/low-emissions-coal/nsw-clean-coal-fund-research-projects#Storage-of-Captured-Carbon-Dioxide-(CO-sub-2–sub-).

[4] http://www.berkelenrodenrijs.net/main.php?lees=3329 (in Dutch).

Norway's Sleipner gas field is the oldest plant that stores CO_2 on an industrial scale. According to an environmental assessment of the gas field that was conducted after 10 years of operation, it was affirmed that geo-sequestration of CO_2 was the most definite form of permanent geological storage of CO_2. Geological information shows absence of major tectonic events after the deposition of the Utsira formation (saline reservoir). This implies that the geological environment is tectonically stable and a site suitable for carbon dioxide storage.

The Weyburn–Midale Carbon Dioxide Project in Weyburn, Saskatchewan, Canada, has determined that the likelihood of stored CO_2 release is less than 1% in 5,000 years.[5] A more recent report, however, claimed evidence of leakage in land above that project.[6]

Detailed geological histories of basins are required. A vast database exists from petroleum seismic data sets. This information could be applied to reduce the risk associated with fault stability. On injection of CO_2 into the earth, there is a major pressure front that can break the seal and make faults unstable. The Gippsland Basin in Australia has a 3D-GEO seismic megavolume that consists of 30+ 3D seismic volumes that have been merged. Such data sets can image faults at a resolution of 15 meters over an area 62 miles (100 km) by 62 miles (100 km). By mid 2010, the first full geological study of the Gippsland Basin became open file by 3D-GEO, making CCS fault risk workflow available with the associated data that constrains it. In other basins around the world, such studies are not available and can only be purchased at high costs.

The Global CCS Institute identified 74 large-scale integrated projects in its 2011 Global Status of CCS report.[7] The following are some examples:

- Salah CO_2 Injection, Northern Africa: This is a fully operational onshore gas field with CO_2 injection. CO_2 is separated from produced gas and reinjected in the producing hydrocarbon reservoir zones. Since 2004, about 1 Mt/a of CO_2 has been captured during natural gas extraction and injected into the Krechba geologic formation at a depth of 1,800 meters.
- Sleipner CO_2 Injection, Norway: This is a fully operational offshore gas field with CO_2 injection. CO_2 is separated from produced gas and reinjected in a saline aquifer above the hydrocarbon reservoir zones.
- Snøhvit CO_2 Injection, Norway: Snøhvit is a fully operational offshore gas field with CO_2 injection. The LNG plant is located onshore. CO_2 is separated from produced gas and injected in a saline aquifer below the H/C reservoir zones offshore. This liquefied

[5] Allan Casey, "Carbon Cemetery," *Canadian Geographic Magazine,* January/February 2008, 61.
[6] Paul Lafleur, "Geochemical Soil Gas Survey, A Site Investigation of SW30-5-13-W2M, Weyburn Field, Saskatchewan," August 27, 2010.
[7] See http://cdn.globalccsinstitute.com/sites/default/files/the_global_status_ccs_2011_2_projects_2.pdf.

natural gas (LNG) plant captures 0.7 Mt/a of CO_2 and injects it into the Tubåen sandstone formation 2,600 meters under the seabed for storage.

- Weyburn Operations, Canada: The oil field is currently operating while injecting CO_2 to increase oil production. This project captures about 2.8 Mt/a of CO_2 from a coal gasification plant located in North Dakota, USA, transports this by pipeline 320 km across the Canadian border, and injects it into depleting oil fields where it is used for enhanced oil recovery.
- Salt Creek Enhanced Oil Recovery, USA: Anadarko Petroleum Corporation will build a pipeline to inject CO_2 in existing Salt Creek Oil Field for enhanced oil recovery. Anadarko has injected 5.12 billion cubic meters of carbon dioxide into the field as part of a project to tease more oil from the field and in the process sequester a greenhouse gas that would otherwise have to be discharged into the atmosphere.
- Enid Fertilizer, USA: The Enid Fertilizer plant sends 675,000 metric tons of CO_2 to be used for EOR. The pipeline and wells are operated separately by Anadarko Petroleum.
- Sharon Ridge EOR, USA: CO_2 from Mitchell, Gray Ranch, Puckett, and Turrell gas processing plants is transported via the Val Verde and CRC pipelines for EOR (including Sharon Ridge EOR field).
- Rangely Weber Sand Unit CO_2 Injection Project, USA: ChevronTexaco, the current owner/operator of the Rangely Weber Sand Unit, has been injecting carbon dioxide into the Rangely Oil Field since 1986 to increase the total volume of recoverable crude oil.

Examples of carbon sequestration at an existing U.S. coal plant can be found at utility company Luminant's pilot version at its Big Brown Steam Electric Station in Fairfield, Texas. This system converts carbon from smokestacks into baking soda. Skyonic plans to circumvent storage problems of liquid CO_2 by storing baking soda in mines, landfills, or simply to be sold as industrial or food-grade baking soda. Green Fuel Technologies is piloting and implementing algae-based carbon capture, circumventing storage issues by then converting algae into fuel or feed.

4.4.2 Recycling and New Markets

CO_2 recycling is a potential response to reducing greenhouse gas emissions from power plants. Technologies under development, such as Bio CCS Algal Synthesis, utilize pre-smokestack CO_2 as a feedstock input to the production of oil-rich algae in solar membranes to produce oil for plastics and transport fuel, and nutritious stock feed for farm animal production. The CO_2 and other captured greenhouse gases are injected into the membranes containing wastewater and select strains of algae causing, together with sunlight or UV light, an oil-rich biomass that doubles in mass every 24 hours.

The Bio CCS Algal Synthesis process is based on earth science photosynthesis. Bio CCS Algal Synthesis test facilities are being piloted at Australia's three largest coal-fired power stations (Tarong, Queensland; Eraring, New South Wales; Loy Yang, Victoria) using piped pre-emission smokestack CO_2 (and other greenhouse gases) as feedstock to grow oil-rich algal biomass in enclosed membranes for the production of plastics, transport fuel, and nutritious animal feed.

Another potential method of addressing CO_2 is to convert it into hydrocarbons, where it can be stored or reused as fuel or to make plastics.

Carbon dioxide scrubbing variants also exist based on potassium carbonate, which can be used to create liquid fuels, although this process requires a great deal of energy.

A proven process to produce a hydrocarbon is to make methanol. Methanol can be synthesized from CO_2 and H_2.

At the department of Industrial Chemistry and Engineering of Materials at the University of Messina, Italy, there is a project to develop a system that works like a fuel-cell in reverse, whereby a catalyst is used that enables sunlight to split water into hydrogen ions and oxygen gas. The ions cross a membrane, where they react with the CO_2 to create hydrocarbons.[8]

The Fischer–Tropsch process may also be used to convert the CO into hydrocarbons. When CO_2 is heated to 2,400°C, it splits the molecule into carbon monoxide (CO) and oxygen. This temperature can be reached by using a chamber containing a mirror to focus sunlight on the gas. Research teams are developing such chambers, at Solarec and at Sandia National Laboratories in New Mexico. According to Sandia, these chambers could provide enough fuel to power 100% of domestic vehicles using 5,800 km^2; unlike biofuels, this would not take fertile land away from crops but would be land that is not being used for anything else.

4.5 Commercial Status

The United States has four synthetic fuel projects, all of which have publicly announced plans to incorporate carbon capture and storage. They are

- American Clean Coal Fuels, known as the Illinois Clean Fuels (ICF) project. This program is developing a 30,000 barrel (4,800 m^3) per day biomass and coal to liquids project in Oakland, Illinois. It plans to market the CO_2 created at the plant for enhanced oil recovery applications. The project is expected to

[8] See David Biello, "Chemical Process Makes Fuel from Carbon Dioxide," *Scientific American,* September 16, 2006, http://www.scientificamerican.com/article.cfm?id=chemical-process-makes-fu.

come online in mid-2013. By combining sequestration and biomass feedstocks, the ICF project will achieve reductions in the life-cycle carbon footprint of the fuels they produce. If sufficient biomass is used, the plant should have the capability to go life-cycle carbon negative, meaning that effectively, for each gallon of its fuel that is used, carbon is pulled out of the air and put into the ground.

- Baard Energy, in its Ohio River Clean Fuels project, is developing a 53,000 barrel (8,400 m^3) per day coal and biomass to liquids project, which has announced plans to market the plant's CO_2 for enhanced oil recovery.
- DKRW is developing a 15,000–20,000 barrel (2,400–3,200 m^3) per day coal to liquids plant in Medicine Bow, Wyoming, which will market its plant's CO_2 for enhanced oil recovery. The project is expected to begin operation in 2013.

In 2009, the U.S. Department of Energy awarded grants to 12 Industrial Carbon Capture and Storage (ICCS) projects to conduct a Phase 1 feasibility study. The DOE plans to select 3–4 of those projects to proceed into Phase 2, design and construction, with operational start-up to occur by 2015. Battelle Memorial Institute, Pacific Northwest Division, Boise, Inc., and Fluor Corporation are studying a CCS system for capture and storage of CO_2 emissions associated with the pulp and paper production industry. The site of the study is the Boise White Paper L.L.C. paper mill located near the township of Wallula in Southeastern Washington state. The plant generates 1.2 MMt of CO_2 annually from a set of three recovery boilers that are mainly fired with black liquor, a recycled byproduct formed during the pulping of wood for paper making. Fluor Corporation will design a customized version of its Econamine Plus carbon capture technology. The Fluor system will be designed to remove residual quantities of remnant air pollutants from stack gases as part of the CO_2 capture process. Battelle is leading the preparation of an Environmental Information Volume (EIV) for the entire project, including geologic storage of the captured CO_2 in deep flood basalt formations that exist in the greater region. The EIV will describe the necessary site characterization work, sequestration system infrastructure, and monitoring program to support permanent sequestration of the CO_2 captured at the plant.

In the United Kingdom, a tender process for a CCS demonstration project. The project will use post-combustion technology on coal-fired power generation at 300–400 MW or equivalent. The project aims to be operational by 2014. The government announced in June 2008 that four companies had prequalified for the following stages of the competition.

Doosan Babcock has modified its Clean Combustion Test Facility (CCTF) in Renfrew, Scotland, to create the largest oxyfuel test facility currently in the world. Oxyfuel firing on pulverized coal with recycled flue gas demonstrates the operation of a full-scale 40 MW burner for use in coal-fired boilers. Sponsors of the project include the UK Department for Business Enterprise and Regulatory Reform (BERR), as well as a group of industrial sponsors and university partners comprising Scottish and Southern Energy (Prime Sponsor), E.ON UK PLC, Drax Power Limited, ScottishPower, EDF Energy, Dong Energy Generation, Air Products Plc (Sponsors), and Imperial College and University of Nottingham (University Partners).

In Beijing, as of 2009, one major power plant is capturing and reselling a small fraction of its CO_2 emissions.

The German industrial area near the city of Spremberg is home to the world's first CCS coal plant. The mini pilot plant is run by an oxyfuel boiler and is also equipped with a flue-gas cleaning facility to remove fly ash and sulfur dioxide. A Swedish company invested 70 million euros in the 2-year project, which began operation September 9, 2008. The power plant, which is rated at 30 MW, is a pilot project to serve as a prototype for future full-scale power plants. Roughly 240 metric tons a day of CO_2 are being trucked 350 km (220 miles), where it will be injected into an empty gas field. For each metric ton of coal burned, 3.6 metric tons of carbon dioxide are produced.

The Federal Resources and Energy Minister Martin Ferguson in Australia opened the first geosequestration project in the southern hemisphere in April 2008. The demonstration plant is near Nirranda South in South Western Victoria. The plant is owned by the Cooperative Research Centre for Greenhouse Gas Technologies (CO2CRC). CO2CRC is a nonprofit research collaboration supported by government and industry. The project has stored and monitored over 65,000 metric tons of carbon dioxide-rich gas that was extracted from a natural gas reservoir via a well, compressed, and piped 2.25 km to a new well. There, the gas was injected into a depleted natural gas reservoir approximately 2 km below the surface. The project has moved to a second stage and is investigating carbon dioxide trapping in a saline aquifer 1,500 meters below the surface. The Otway Project is a research and demonstration project, focused on comprehensive monitoring and verification.

Development of commercially viable processes for capturing CO_2 from power plants is being funded by the U.S. DOE, electric utility companies, and other organizations. These processes typically use solvents, solid sorbents, and membrane-based technologies for separating and capturing CO_2. Amine-based solvent systems are in commercial use for scrubbing CO_2 from industrial flue gases and process gases. However, solvents have yet to be applied to removing the large volumes of CO_2 that would be required for a coal-fired EGU. Solid sorbents can be used to capture CO_2 through chemical adsorption, physical adsorption, or a combination of the two effects.

Membrane-based capture uses permeable or semipermeable materials that allow for the selective transport/separation of CO_2. Oxy-combustion uses high-purity oxygen (O_2) instead of air to combust coal, producing a highly concentrated CO_2 stream that does not require a separation/capture step.

As noted, once the CO_2 is captured, it is transported, if necessary, and stored. Geologic formations such as oil and gas reservoirs, unmineable coal seams, and underground saline formations are potential options for long-term storage. Basalt formations and organic-rich shales are also being investigated for potential use as storage. Beneficial reuse (e.g., enhanced oil recovery or carbonation) is a potential alternative to strict storage that provides potential revenue to offset a portion of the CCS costs.

A recent study prepared for the U.S. DOE by the Pacific Northwest National Laboratory[9] evaluated the development status of various CCS technologies. The study addressed the availability of capture processes; transportation options (CO_2 pipelines); injection technologies; and measurement, verification, and monitoring technologies. The study concluded that, in general, CCS is technically viable today. However, full-scale carbon separation and capture systems have not yet been installed and fully integrated at electric-generating units. The study also did not address the cost or energy requirements of implementing CCS technology.

In 2010, an Interagency Task Force on Carbon Capture and Storage was established to develop a comprehensive and coordinated federal strategy to speed the commercial development and deployment of CCS technologies. The task force is specifically charged with proposing a plan to overcome the barriers to the widespread, cost-effective deployment of CCS within 10 years, with a goal of bringing 5–10 commercial demonstration projects online by 2016. As part of its work, the task force prepared a report that summarizes the state of CCS and identified technical and nontechnical barriers to implementation.

4.6 Improving Efficiencies

A range of efficiency improvements can be applied to coal-fired plants to increase thermal efficiency of power production and thereby reduce GHG

[9] Pacific Northwest National Laboratory (PNNL). *An Assessment of the Commercial Availability of Carbon Dioxide Capture and Storage Technologies as of June 2009*, PNNL-18520. Pacific Northwest National Laboratory, Richland, WA. June 2009. http://www.pnl.gov/main/publications/external/technical_reports/PNNL-18520.pdf.

[10] James Graham, "Microwaves for Coal Quality Improvement: The Drycol Project," DBAGlobal Australia, Milton Queensland, Australia. Presented at the SACPS/International Pittsburgh Coal Conference 2007, Johannesburg, South Africa, September 10–14, 2007. Available at http://www.drycol.com/downloads/Drycol%20Paper%20ACPS-1%20060608.pdf.

emissions. Following is a summary of options and reported efficiency improvements gained:

- Combustion Control Optimization—Combustion controls adjust coal and air flow to optimize steam production for the steam turbine/generator set. However, combustion control for a coal-fired EGU is complex and impacts a number of important operating parameters, including combustion efficiency, steam temperature, furnace slagging and fouling, and NOx formation. The technologies include instruments that measure carbon levels in ash, coal flow rates, air flow rates, CO levels, oxygen levels, slag deposits, and burner metrics as well as advanced coal nozzles and plasma-assisted coal combustion. Reported efficiency gains: 0.15% to 0.84%.
- Cooling System Heat Loss Recovery—Recover a portion of the heat loss from the warm cooling water exiting the steam condenser prior to its circulation through a cooling tower or discharge to a water body. The identified technologies include replacing the cooling tower fill (heat transfer surface) and tuning the cooling tower and condenser. Reported gains: 0.2% to 1%.
- Flue-Gas Heat Recovery—Flue-gas exit temperature from the air preheater can range from 250°F to 350°F depending on the acid dew point temperature of the flue gas, which is dependent on the concentration of vapor phase sulfuric acid and moisture. For power plants equipped with wet FGD systems, the flue gas is further cooled to approximately 125°F as it is sprayed with the FGD reagent slurry. However, it may be possible to recover some of this lost energy in the flue gas to preheat boiler feedwater via use of a condensing heat exchanger. Reported gains: 0.3% to 1.5%.
- Low-Rank Coal Drying—Subbituminous and lignite coals contain relatively large amounts of moisture (15% to 40%) compared to bituminous coal (less than 10%). A significant amount of the heat released during combustion of low-rank coals is used to evaporate this moisture rather than generate steam for the turbine. As a result, boiler efficiency is typically lower for plants burning low-rank coal. The technologies include using waste heat from the flue gas and/or cooling water systems to dry low-rank coal prior to combustion. Reported gains: 0.1% to 1.7%.
- Soot Blower Optimization—Soot blowers intermittently inject high-velocity jets of steam or air to clean coal ash deposits from boiler tube surfaces in order to maintain adequate heat transfer. Proper control of the timing and intensity of individual soot blowers is important to maintain steam temperature and boiler efficiency. The identified technologies include intelligent or neural-network soot blowing (i.e., soot blowing in response to real-time conditions in the boiler) and detonation soot blowing. Reported gains: 0.1% to 0.65%.
- Steam Turbine Design—There are recoverable energy losses that result from the mechanical design or physical condition of the steam turbine. For example, steam turbine manufacturers have improved the design of turbine blades and steam seals, which can increase both efficiency and output (i.e., steam turbine dense pack technology). Reported gains: 0.84% to 2.6%.

Table 4.3 provides examples and summaries of technologies.

Table 4.3 Efficiency Improvement Technologies and Examples

Efficiency Approach	Technology Description	Examples
Coal Drying	Low-rank coals (lignite and subbituminous) are utilized because the low cost per unit of heat input relative to bituminous coal and the low sulfur content. However, a major disadvantage of low-rank coals is their high moisture content, typically 25% to 40%. When this coal is burned, considerable energy is required to vaporize and heat the moisture, thus raising the heat rate of the EGU and lowering its efficiency. As fuel moisture decreases, the heating value of the fuel increases so that less coal needs to be fired to produce the same amount of electric power. Drier coal is also easier to handle, convey, and pulverize, reducing the burden on the coal-handling system. In addition, a boiler designed for dried coal is smaller and has lower capital costs than a comparable plant designed to burn coal that has not been dried. The pre-combustion drying of low-rank coals can improve the overall efficiency, and several advanced coal-drying technologies are or nearly are commercially available.	Great River Energy developed a coal-drying technology for low-rank coals in partnership with the U.S. DOE as part of the DOE's Clean Coal Power Initiative. The technology has been successfully demonstrated on a PC-fired boiler burning lignite at the utility's Coal Creek Station in Underwood, North Dakota. The technology is now commercially offered under the trade name DryFining. The DryFining process passes warm cooling water from the steam turbine exhaust condenser through an air heater where ambient air is heated before being sent to a fluidized-bed coal dryer. The dried coal leaving the fluidized bed is sent to a pulverizer and then to the boiler. Air leaving the fluidized bed is filtered before being vented to the atmosphere. In addition to using power plant waste heat to reduce moisture. DryFining also segregates particles by density. This means a significant amount of higher density compounds containing sulfur and mercury can be sorted out and returned to the mine rather than utilized in the boiler. The end result is that more energy can be extracted from the coal while simultaneously reducing emissions of mercury, sulfur dioxide, and NOx. At the Coal Creek Station, the process increased the energy content of the lignite from 6,200 to 7,100 Btu/lb, thereby resulting in a decrease in the fuel input into the boilers by 4% and a corresponding decrease in CO_2 emissions. Net gains in overall efficiency of 2% to 4% are reported for the process.

RWE Power in Germany is also developing a fluidized-bed drying technology for lignite, called WTA. A difference between this and the preceding drying processes is the WTA process first mills and then dries the lignite, whereas the DryFining process first dries and then mills the lignite. A prototype commercial-scale drying plant using the WTA process began operation in 2009 at the utility's Nederaussem Power Station site. For the WTA process, lignite is first milled to a fine particle size by hammer mills in direct series with a two-stage fluidized-bed dryer. The dried fuel exiting the stationary bed is separated from the gas stream and mixed with coarser lignite solids collected from the bottom of the dryer bed and then fed directly to the boiler. The heat needed for the drying of the fuel is supplied by external steam, which is normally taken from the turbine with the heat transfer taking place in tube bundles located inside the bed. Based on the development work completed to date of the WTA technology, the net gain in cycle efficiency is reported to be on the order of 4 percentage points, depending on the moisture content of the raw coal and the final moisture of the dried lignite.

One coal drying process being developed by DBAGlobal Australia Pty, Ltd., with the trade name Drycol process uses the controlled application of microwave radiation to dry coal.[10] Coal feedstock is first separated into fine grade coal and one or more larger grades. The fine coal is loaded onto a conveyor and conveyed continuously through a microwave-energized heating chamber for

(Continued)

Table 4.3 Efficiency Improvement Technologies and Examples—*cont'd*

Efficiency Approach	Technology Description	Examples
		drying. The fine grade coal is dried sufficiently so that when it is recombined with the larger grade coals, the moisture content of the aggregate coal is within a target moisture content range. Other coal-drying technologies for low-rank coals in various stages of development include (1) attrition milling of coal followed by air drying to produce a low-moisture coal product; (2) compressing heated, coarse-crushed coal to squeeze out water; and (3) heating wet coal under pressure to approximately 480°F to 570°F.
Boiler Feedwater Heating and Hot-Windbox	The high-pressure liquid water entering the steam generator is called feedwater. A feedwater heater is a component used to preheat water delivered to the boiler section. Thermodynamic optimization of this cycle is important to overall efficiency. In a conventional plant, the energy used to heat the feedwater is steam extracted between the stages of the steam turbine. Therefore, approximately a quarter of the steam that would be used to perform expansion work in the turbine (and generate power) is not utilized for that purpose. However, using other heat sources for the feedwater heater avoids the need to extract steam from the turbine, allowing the steam to be used for electric power generation, and increases the output of the steam cycle and potentially lowers GHG emissions. This alternate heat source can either be from an integrated solar thermal energy source or from a combustion turbine.	Examples of solar thermal energy used to augment the steam cycle at combined cycle facilities include the Martin Next Generation Solar Center in Florida and the proposed Green Energy Partners/Stonewall, LLC, facility in Virginia. The first coal-fired power plant to integrate solar thermal technology is the Cameo Generating Station in Colorado. In addition, the Electric Power Research Institute (EPRI) is currently evaluating adding solar thermal energy to the Escalante and Mayo coal-fired power plants. An example of combustion turbine integration for feedwater heating is the Kettle Falls Generating Station. For coal-fired boiler systems optimized to accommodate the combustion turbine exhaust, the incremental fuel efficiencies would be expected be comparable with combined cycle generation.

		Another potential approach to integrate the use of a combustion turbine with a coal-fired steam cycle is using the turbine exhaust directly in the boiler in a hot-windbox. This involves injection of the combustion turbine exhaust directly into the boiler windbox or primary air ducts to provide an oxygen source as well as a heat source.
Combined Heat and Power Plants	Coal-fired plants dedicated to electric power generation and using the latest commercially available advanced technologies will generally operate at overall net efficiencies of approximately 40%. Significant amounts of energy released by coal combustion are lost during the steam condensation segment of the Rankine cycle due to heat transfer into the cooling water. In Europe, electricity is commonly generated by facilities that serve as both electricity generators and thermal energy producers for the local town or city district heating system. These combined heat and power (CHP) facilities are also known as cogeneration facilities. Operating an electric power station in a CHP mode allows recovery of some of the heat that would otherwise be rejected into cooling water, improving the overall efficiency of energy utilization. In applying CHP to an existing or new facility, the temperature of the cooling water is normally not high enough to meet the requirements for most district heating or industrial process applications. In these cases, steam would be extracted at an elevated pressure and temperature from an intermediate stage of the steam turbine and then used for district or process heating. This results in a decrease in the total electric power generation from the facility. However, the overall fuel efficiency of CHP is higher than if electricity and steam were	DuPont's titanium dioxide plant in Johnsonville, Tennessee, is located next to TVA's Johnsonville power plant and buys high-pressure process steam from the 1,200 MW facility. The power plant is composed of 10 coal-fired boiler steam turbine units; the DuPont plant uses steam extracted from Units 1 through 4. Providing steam to the DuPont facility at the required process pressures reduces overall output of the power plant by 50 MW.

(*Continued*)

Table 4.3 Efficiency Improvement Technologies and Examples—*cont'd*

Efficiency Approach	Technology Description	Examples
	generated separately. Because electricity can be transmitted over long distances, electric power plants can be located in remote areas as well as urban areas. However, thermal energy cannot be effectively transported over extended distances. This limits the practicality of incorporating a CHP mode into many electric power plant designs. The EGU needs to be located in close proximity to either a district energy system or an industrial facility with a significant and steady thermal demand.	
		Blue Flint Ethanol in Underwood, North Dakota, is a 50 million gallon per year dry mill ethanol producer located next to Great River Energy's 1,160 MW Coal Creek lignite-fired power plant. Starting operations in 2007, the ethanol facility purchases approximately 100,000 pounds per hour of medium-pressure steam extracted from the power plant. Goodland Energy Center in Goodland, Kansas, is a small 22 MW coal-fired power plant that is supplying steam to a 20 million gallon per year ethanol plant and a 12 million gallon per year biodiesel plant, both co-located with the power plant. The power plant and the ethanol plant both started construction in 2006.
Oxygen Combustion	The basic concept of oxy-combustion is to use oxygen-enriched air and recycled flue gas (containing mostly CO_2) in place of ambient air for coal combustion. The resulting flue gas contains primarily CO_2 and water vapor with smaller amounts of oxygen, nitrogen, SO_2, and NOx. Consequently, the flue gas can be processed relatively easily to further purify the CO_2 (if necessary)	A 30 MW oxy-firing pilot plant at the Schwarze Pumpe station in Spremberg, Germany. This plant is the first complete oxy-combustion unit that includes the integrated system from the air separation unit to the gas purification and compression systems. The CO_2 will be compressed and liquefied for storage experiments to be conducted.

for use in enhanced oil or gas recovery or for geological storage. An oxy-combustion power plant consists of an air separation unit (ASU), an electric-generating unit with O_2-blown combustion, and a CO_2 treatment unit. The conventional ASU is a cryogenic process that has a significant energy requirement. However, alternative oxygen separation methods are being researched for possible commercial scale development. These alternative methods include ion transport membranes (ITM), ceramic autothermal recovery, oxygen transport membranes, and chemical looping. Oxygen is mixed with recirculated flue gas to create a mixture of O_2 and CO_2 (and some H_2O), which is used as the source of combustion oxidant instead of ambient air. The absence of air nitrogen produces a flue-gas stream with a high concentration of CO_2.	A 32 MW oxy-firing demonstration project in France retrofitting an existing boiler to natural gas oxy-combustion. The captured CO_2 will be transported through an approximately 19-mile long pipeline and stored in a depleted gas field in Lacq, in the south of France. A comprehensive test program using the 15 MW tangentially fired Boiler Simulation Facility and 15 MW Industrial Scale Test Facility operated by Alstom Power, Inc., in Windsor, Connecticut. Testing is being conducted to assess a broad range of oxy-combustion design options. Project partners include the U.S. DOE, the Illinois Clean Coal Institute, and 10 electric utility companies.

5 The TVA Ash Spill

Chapter Outline

5.1 Introduction

This chapter examines the December 2008 ash pond spill from the Tennessee Valley Authority's (TVA) Kingston coal-fired plant. This was the largest industrial spill ever to have occurred on the North American continent. An examination of the conditions from the aftermath of the spill led to several conclusions:

- PM_{10} concentrations measured in the immediate community were influenced by the facility site conditions. Measurements made at various locations in the immediate community neighboring the TVA site were shown to be closely correlated with PM measurements obtained from within the TVA property and influenced by on-site activities performed by TVA and its contractors.
- There was no evidence to support that the PM_{10} National Ambient Air Quality Standard (NAAQS) had been exceeded. A database of nearly 268,000 PM_{10} measurements does not support excessive dust emissions occurred in the community that exceeded the NAAQS annual or 24-hour guidelines.
- Although PM_{10} values did not exceed NAAQS, fly ash was nonetheless released into the community. Fly ash contains a broad array of toxins, including but not limited to arsenic, barium, beryllium, boron, cadmium, chromium, chromium VI,

Handbook of Pollution Prevention and Cleaner Production Vol. 4. DOI: http://dx.doi.org/10.1016/B978-1-4377-7815-1.00005-9

cobalt, copper, fluorine, lead, manganese, nickel, selenium, strontium, thallium, vanadium, zinc, and other hazardous constituents. Based on the literature of reported compositions of these toxins found in coal ash combustion products (fly ash), the average composition of each of these constituents was calculated using the mean and maximum PM_{10} concentrations for the baselines recorded for each general site location. These concentrations were compared against the safe air quality guidelines recommended by the World Health Organization (WHO). The analysis supports that the community was exposed to arsenic, cadmium, chromium (total and valence VI), lead, manganese, nickel, and vanadium at levels that either exceeded the WHO guidelines (in some instances) or were above national averages for rural areas.

5.2 Facility Background

The Tennessee Valley Authority's Kingston Fossil Plant is a coal-fired electricity-generating plant. In this plant, pulverized coal is burned in boilers to heat water to about 1,000°F (540°C) to create high-pressure steam. The steam is exhausted into turbines at pressures of more than 1,800 pounds per square inch (130 kilograms per square centimeter). The turbines are connected to generators that spin at 3,600 revolutions per minute to make alternating current electricity at 20,000 volts. River water is pumped through tubes in a condenser to cool and condense the steam coming out of the turbines. Kingston generates about 10 billion kilowatt-hours of electricity a year, enough to power about 670,000 homes.

Coal is a compact, stratified mass of metamorphosed plants that have stopped decaying to varying degrees.[1] It consists of primarily carbon (C) and lesser amounts of ash, moisture, hydrogen (H_2), oxygen (O_2), sulfur (S), and nitrogen (N_2). It also contains small quantities of uranium (238U and 235U) and thorium (232th).[2] Over 95% of ash content of coal is made up of alumina ($Al2O_3$), silica (SiO_2), lime (CaO), iron oxide (Fe_2O_3), and magnesia (MgO). The remainder is made up of sodium oxide (Na_2O), potassium oxide (K_2O), phosphorus pentoxide (P_2O_5), chlorine (Cl), and titanium oxide (TiO_2), and trace amounts of other chemicals.[3]

The burning of coal results in the production of combustion gases, which are the result of reactions of the chemical elements (i.e., H_2, C, S, and N_2) with oxygen and ash. The combustion gases leaving the boiler must be cleaned to

[1] H.H. Lower, ed., *Chemistry of Coal Utilization,* vol. 1, New York: John Wiley & Sons, 1945, 2.

[2] J.P. McBirds, R.E. Moore, J.P. Witherspoon, and R.E. Blanco, "Radiological Impact of Airborne Effluents of Coal-Fired and Nuclear Power Plants," ORNL 5315, Oak Ridge National Laboratory, August 1977.

[3] Lower, *Chemistry of Coal Utilization,* 486.

remove acid gas precursors such as oxides of sulfur and nitrogen and fine ash particles known as fly ash. The other coal combustion product is bottom ash—the heavier ash particles that are collected at and removed from the bottom of the boiler generally as solid material. The fly ash is collected using solid removal equipment and piped to ash ponds in slurry form. The bottom ash is landfilled.

The Kingston plant consists of nine units. All nine units are pulverized coal units designed by Combustion Engineering. Units 1–4 have a nameplate or design capacity of 175 MW, and units 5–9 have design capacities of 200 MW each, for a total capacity of 1,700 MW. Units 1–4 are natural circulation, reheat-type radiant boilers; units 5–9 are twin furnace, reheat-type with controlled circulation. Construction of these units began in the early 1950s and was completed by 1956. At the time of the incident, the Kingston Plant was one of the largest coal-fired power plants in the world.

As of December 2008, the nine units had been in continuous service for 53 to 55 years with a combined cumulative 485 years of service. Table 5.1 provides the commission dates and nameplate ratings of the nine units.

Because of fluctuations in the average atmospheric temperature, the humidity, and other factors, the actual plant output during winter and summer seasons is different from the design capacity. The plant capacity is 1,425 MW in summer and 1,447 MW in winter, resulting in an annual average capacity of 1,435 MW or a maximum generating capacity of 12.6 (or $1{,}435 \times 365 \times 24$) billion kilowatts hours per year. As noted previously, TVA's website states, "Kingston generates about 10 billion kilowatt-hours of electricity a year."[4] In other words, this plant operates 80% of the time or has 80% availability.

Combustion or burning of coal produces sulfur oxides (SOx), nitrogen oxides (NOx), and fly ash. These pollutants are hazardous and can impact human health, vegetation, and wild life and are removed from the combustion gaseous products or flue gas. To remove fly ash and reduce chimney opacity, TVA installed electrostatic precipitators (ESPs) at the Kingston Plant in 1978.[5]

Figure 5.1 shows the flue-gas treatment train after installation of the ESPs. Units 1–4 are arranged such that the flue gas exits the precipitators into ducts that lead to two induced draft fans per unit. Flue gas discharges from the fans into a common plenum and then into a 1,000-foot stack before exhausting into the atmosphere. Flue gas from unit 5 exits the precipitator into two ducts that are connected to the same common plenum and stack as units 1–4. Units 6–9 have an arrangement exactly like unit 5 except that they discharge into atmosphere through a separate common plenum and 1,000-foot stack.

[4] http://www.tva.com/sites/kingston.htm

[5] "Final Environmental Assessment," TVA, April 2006.

Table 5.1 TVA Kingston Plant Coal Type, Number of Units, Capacity, and Age

Unit No.	Nameplate Capacity (Megawatts)	Summer Capacity (Megawatts)	Winter Capacity (Megawatts)	Coal Type No.1	Coal Type No. 2	Initial Operation Date	Unit Age As of December 2008 (Year–Month)
1	175	135	138	Bituminous	Subbituminous	Feb-54	54–10
2	175	135	138	Bituminous	Subbituminous	Apr-54	54–8
3	175	135	138	Bituminous	Subbituminous	Jun-54	54–6
4	175	135	138	Bituminous	Subbituminous	Jul-54	54–5
5	200	177	179	Bituminous	Subbituminous	Jan-55	53–11
6	200	177	179	Bituminous	Subbituminous	Mar-55	53–9
7	200	177	179	Bituminous	Subbituminous	May-55	53–7
8	200	177	179	Bituminous	Subbituminous	Sep-55	53–4
9	200	177	179	Bituminous	Subbituminous	Dec-55	53–0
Total	1700	1425	1447	-			485 Years + 2 Months

Source: http://www.eia.doe.gov/cneaf/electricity/page/capacity/capacity.html

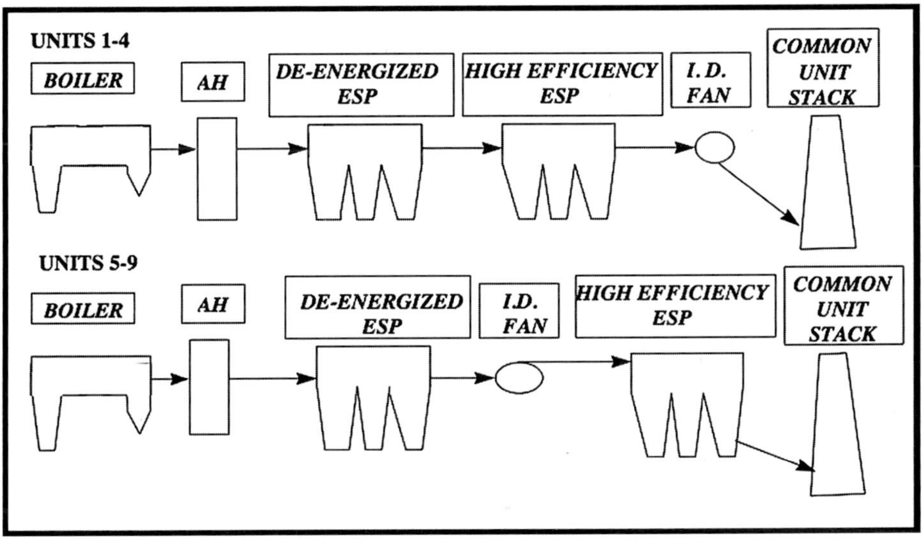

Figure 5.1 Flue-gas treatment train after installing ESP at Kingston Plant.
Source: Draft Environmental Assessment, TVA, April 2002.

Guillotine dampers are located in each duct just upstream of the common plenums for all units to allow the isolation of an individual unit.[6]

Two 1,000-foot chimneys replaced the original nine chimneys in 1976 (units 1–4 had chimneys 250 feet tall each, and units 5–9 had chimneys that were 300 feet tall each). Eight selective catalytic reduction (SRC) systems were put into service in 2004 to reduce NOx emissions. A ninth system was brought online in 2006. These systems are installed at the high dust stream between the boiler and ESP. The photograph in Figure 5.2 was downloaded from the Internet showing the two 1,000-foot stacks.

To meet SO_2 emission limits, the Kingston Plant switched to low-sulfur coal. The sulfur content of the coal used at the plant ranged from 1.2 to 2.3 pounds (lb). SO_2/million British thermal units (MMBtu) from 1978 to 2000. The plant operated at or below 1.5 lb SO_2/MMBtu from 2000 through 2005. The State Implementation Plan's (SIP) SO_2 limit for the plant was 2.8 lb SO_2/MMBtu in 2006. To meet future SO_2 emission limits, TVA has been constructing two scrubbers to process flue gas from all nine boilers at the plant. The scrubbers are installed downstream of the current particulate and NOx control systems[7] and began operation in late 2010.

[6] Donald Borio and Robert Babb, "Technical and Economic Considerations in Hot and Cold Placement of SCR Systems for Utility Boilers," ICAC Forum '02. http://www.icac.com/Files/Borio.pdf.

[7] "Final Environmental Assessment," TVA, April 2006.

Figure 5.2 Photo of the 1,000-foot-tall stacks at the Kingston Plant.

5.3 Coal Sources and Quantities

The Kingston Plant is located near the Southern Appalachian coal-producing region of the United States. This coal-mining region includes mines of eastern Kentucky, Virginia, West Virginia, Tennessee, Northern Alabama, and Northern Georgia. These mines generally produce low- to mid-sulfur coal demanding a premium price, especially since enactment of the Clean Air Act amendment of 1990.

The Kingston Plant received about 3.8 to 4.0 million tons of medium sulfur coal from Eastern Kentucky annually until the early 2000s.[8,9] In early 2000, TVA modified the plant by installing the EPRICON process—a flue-gas conditioning process developed by EPRI allowing continued use of ESPs that were originally designed for lower ash content than ash content of low-sulfur coals—and switched to low-sulfur coal to meet sulfur emission limits. Since about 2000, the **Kingston** plant has burned 4.0 to 4.2 million tons per year of low-sulfur coal from Virginia, West Virginia, and Powder River, Wyoming.[10,11] In 2010, TVA began operating two scrubbers at Kingston Plant,

[8] "Draft Environmental Assessment," TVA, April 2002.

[9] "Energy Vision 2020," Technical Document 1, TVA, December 1995.

[10] http://www.tva.gov/power/coalart.htm, accessed November 16, 2010.

[11] Dave Flessner, "Tennessee: Low-Sulfur Coal Price Doubles, Following Gas & Oil," November 30, 2004, http://www.energybulletin.net/node/3439?ENERGYBULL=70e0a3702c92c758ac2a4926e2f921d6.

allowing it to burn higher sulfur content, less expensive coal while meeting SO_2 emission limits.

5.4 Solid Waste Storage

Historically, Kingston Plant has been generating two solid waste streams: bottom ash collected from the boilers and fly ash collected in ESPs. Generally, about 20% of the coal ash is collected as bottom ash and 80% as fly ash. Particulate capture devices do not capture 100% of the fly ash, and very fine particles are carried over by the flue gas into the atmosphere through the stacks. TVA reported using over 3.9 million tons of coal, producing over 0.28 million tons of fly ash and 0.7 million tons of bottom ash in 1994.[12] Assuming a coal ash content of 5%–9%, which is typical of the Appalachian coal and an annual coal consumption of 3.8–4.2 million tons per year, the Kingston Plant has produced 0.19 to over 0.38 million tons of ash annually since it began operation in 1954. Over the period of 1954 to 2009, the plant has generated between 8 and 16 million tons of fly ash as well as 2 to 4 million tons of bottom ash. All of this waste is disposed of on-site.

An initial 85-acre ash pond was completed for ash disposal in 1954 followed by completion of a 275-acre storage area in 1958. The 275-acre storage area consisted of the main pond where ash slurry was pumped for storage and a stilling pond where fine ash particles settle out and the decanted water is sent back to the reservoir.

The initial ash ponds were full by the early 1980s. To create additional ash storage area, TVA began to dredge ash from a collection pond and to construct elevated dredge cells with containment dikes where the ash from the main pond was finally placed. Between 1954 and 1990, the loose wet ash placed in the dredge cells accumulated to a depth of 40 to 65 feet.[13] Figure 5.3 shows an aerial photograph of the site.

Prior to 1958, before Dike C was completed to fully contain the ash pond, the ash stream was discharged directly to the Watts Bar Reservoir. Modifications and additions to increase the original storage capacity of the ash ponds were made throughout the 1970s. In the mid-1980s, Dike C was raised for a third and final time.

According to the published reports, in 1995, TVA began construction of a vertical dredge cell expansion program. A Solid Waste Permit was issued by the Tennessee Department of Environment and Conservation (TDEC) to raise

[12] Figure T1-74, Energy Vision 2020, Technical Document 1, TVA, December 1995.

[13] "Root Cause Analysis of TVA Kingston Dredge Failure from December 22, 2008," AECOM, Project No. 60095742, June 25, 2009.

Figure 5.3 Aerial view of the ash storage area

Dredge Cells 1, 2, and 3 to form an approximately 120-acre sluiced ash storage structure. The raising program of the cells was designed in 1995 and was planned for completion by 2014.

Additional plans for extending the life of the storage area beyond 2014 were prepared between 2004 and 2006. TVA reported that 100,000 cubic yards of ash material were placed within the 31-acre Dredge Cell 2 in two months—from October 16 to December 18, 2008. AECOM estimated vertical fill height rate is approximately 6 feet per year.

5.5 Ash Analysis

The composition of fly ash varies, but all fly ash includes substantial amounts of silicon dioxide (SiO_2; both amorphous and crystalline); aluminum oxide (Al_2O_3); iron oxide (Fe_2O_3); sulfur trioxide (SO_3); calcium oxide (CaO); and various toxic heavy metals such as arsenic, beryllium, boron, cadmium, chromium, cobalt, lead, manganese, mercury, molybdenum, selenium, strontium, thallium, and vanadium in trace amounts. Table 5.2 shows typical characteristic of coal fly ash.[14] An analysis of ash from the Kingston Plant has revealed high levels of arsenic (As): 75 mg/kg; mercury (Hg): 150 μg/kg; and radioactivity (226Ra + 228Ra): 8 pCi/g.[15] Suspended in the air, the ambient fine particles (<10 μm) containing these toxins may pose a health risk to local communities. Leaching of contaminants from the coal ash can also cause contamination of surface and ground waters.

5.6 The Spill

On December 22, 2008, at approximately 01:00 hours, the northeastern dike at the TVA Kingston Power Plant, located in Harriman, Roane County, Tennessee, failed. The dike retained one of three cells at the facility used for dewatering fly ash. Approximately 5.4 million cubic yards of fly ash were released into two sloughs that flow into the Emory River. The release extended approximately 300 acres outside the ash storage areas. Local emergency officials first responded to the scene and then shortly thereafter began to assist residents affected by the fly ash flow. Three residential homes became

[14] B.B. Maruthi Sridhar and Robert K. Vincent, "Mapping and Estimation of Phosphorus and Copper Concentrations in Fly Ash Spill Area Using LANDSAT TM Images," *Photogrammetric Engineering & Remote Sensing,* September 2009.

[15] Laura Ruhl, et al., "Survey of the Potential Environmental and Health Impacts in the Immediate Aftermath of the Coal Ash Spill in Kingston, Tennessee," *Environmental Science & Technology,* 2009, 43: 6326–6333.

Table 5.2 Reported Literature Composition of Coal Combustion Product Known as Fly Ash

Elements	Concentration Range	Value Relied On
	Major Components (%)	
Alumina	0.1–17.3	
Calcium	0.11–22.2	
Iron	1–29	
Magnesium	0.04–7.5	
Sodium	0.01–2.03	
Potassium	0.15–3.5	
Sulfur	0.002–0.004	
Phosphorus	0.04–0.8	
	Trace Components (ppm)	
Arsenic	2.3–6300	75
Barium	806	806
Beryllium	5	5
Boron	10–618	628
Cadmium	0.7–130	65
Cobalt	7–520	264
Copper	112	112
Chromium	10–1000	505
Chromium VI	90	90
Fluorine	29	29
Lead	3.1–5000	2,502
Mercury	0.02–1	0.15
Manganese	58–3000	1,529
Nickel	6.3–4300	2,153
Selenium	0.2–134	45
Strontium	775	775
Thallium	9	9
Vanadium	252	252
Zinc	10–3500	1,755

Sources: B.B. Maruthi Sridhar and Robert K. Vincent, "Mapping and Estimation of Phosphorus and Copper Concentrations in Fly Ash Spill Area Using LANDSAT TM Images," *Photogrammetric Engineering & Remote Sensing,* September 2009; Laura Ruhl, et al., "Survey of the Potential Environmental and Health Impacts in the Immediate Aftermath of the Coal Ash Spill in Kingston, Tennessee," *Environmental Science & Technology,* 2009, 43: 6326–6333; "Managing Coal Combustion Residues in Mines," Committee on Mine Placement of Coal Combustion Wastes, National Research Council of the National Academies, 2006.

condemned as a result of the release. On December 22, 2008, the National Response Center (NRC) and subsequently the U.S. Environmental Protection Agency (EPA) Region 4 were notified of the incident. A Federal On-Scene Coordinator (OSC) and contractor were mobilized to the TVA Kingston Power Plant facility the same day. Some photographs showing the extent of the spill during a site visit made are shown in Figures 5.4 through 5.9.

The TVA and Environmental Protection Agency initially estimated that the spill released 1.7 million cubic yards (1.3 million m^3) of sludge. After an aerial survey, the official estimate was more than tripled to 5.4 million cubic yards (4 million m^3) on December 25, 2008. The spill covered surrounding land with up to 6 feet (1.8 meters) of sludge.

The 84-acre (0.34 km^2) unlined aboveground ash fill, which was situated 60 feet (18 meters) above the ash pond and 74 feet (23 meters) above the Emory River winter level, 741 feet (226 meters) above sea level, contained a watery slurry of fly ash generated by the burning of finely ground coal at the steam power plant. Fly ash is the fine particulate pollutant produced by the combustion of coal, which is collected rather than being allowed to escape into the atmosphere; it is then mixed with water so it can be pumped into the retaining pond. Once the particulate matter settles out, it is dredged to drying cells. The dredge cell was surrounded by 60-foot (18 meters) tall earthen walls, which had twice developed leaks since 2002.

The spill caused a mudflow wave of water and ash that covered 12 homes, pushing one entirely off its foundation, rendering 3 homes uninhabitable and causing damage to 42 residential properties. It also washed out a road, ruptured a major gas line, obstructed a rail line, downed trees, broke a water main, and destroyed power lines. Initially, 22 residences were evacuated. It

Figure 5.4 Rings of fly ash washing up on dock posts.

Figure 5.5 Collapsed pond and attempts by TVA contractors to initiate cleanup.

Figure 5.6 Ash spill from collapsed dam.

Figure 5.7 Wider area impacted from spill.

was the largest coal-related slurry spill in United States history, more than three times the size of the Martin County sludge spill of 2000, which spilled 306 million U.S. gallons (1.2 million m^3) of liquid coal waste. For comparative purposes, the volume released was about 101 times larger than the 1989 Exxon Valdez oil spill (1.1 billion/10.9 million = 100.9) and 5.4 times larger than the BP Mocando well oil spill in the Gulf of Mexico (1.1 billion/205 million = 5.4).

On December 23, 2008, a TVA spokesman stated that, at the time of the spill, the area contained about 2.6 million cubic yards (2.2 million m^3) of ash, and that two-thirds of that had been released, which would later be found to cover an area of 300 acres (1.2 km^2). The *New York Times* noted that the amount spilled is larger than the amount stated to have been in the pond before the spill, a discrepancy that TVA was unable to explain. The containment area affected was one of three; the other two stayed intact, while only the retaining wall for the 84-acre (0.34 km^2) solid waste containment area was affected.

The spill was reported in various news media to have killed a "tremendous" number of fish. Independent test results, conducted at the Environmental Toxicology and Chemistry laboratories at Appalachian State University, showed

Figure 5.8 Aerial shot of impacted area taken by author.

significantly elevated levels of toxic metals (including arsenic, copper, barium, cadmium, chromium, lead, mercury, nickel, and thallium) in samples of slurry and river water. Limited sampling of surface samples and water obtained from the site by the investigator also confirmed elevated levels of these toxins.

5.7 Facility Environmental Footprint

Ignoring the 2008 spill incident, the environmental footprint of the Kingston coal-burning plant is enormous. According to the EPA's 2002 National Emissions Inventory (NEI), the facility releases the following amounts of pollutants to air on an annual basis:

- 243,600 lbs. VOCs
- 155,142,000 lbs. SO_2
- 122,431 lbs. NH_3
- 2,031,400 lbs. CO
- 52,169,200 lbs. NH_3
- 7,255,529 lbs. Primary PM_{10}

Figure 5.9 Another aerial shot showing extent of spill.

In total, the facility releases 216,964,161 lbs. (pounds) of chemical compounds into the atmosphere each year.

Volatile organic compounds (VOCs), sulfur dioxide (SO_2), and ammonia (NH_3) are recognized by the U.S. EPA as precursors that contribute to the formation of particulate matter (PM). According to the NEI database, 100% of these chemical toxins are released from two 1,000-foot-tall stacks at an average stack gas velocity of 107 feet per second (fps). These tall stacks are intended to disperse the emissions into the upper atmosphere and hence have impacts on the regional air quality; however, these releases would not be expected to have an impact on the immediate vicinity (i.e., the valley region).

The facility does report fugitive emissions of PM. Of the 7,255,529 lbs. of Primary PM_{10} reported to be released, the NEI reports 107,329 lbs. of this is released as fugitive emissions (roughly 1.5% of PM releases are fugitive emissions). Assuming that the 2002 NEI is representative of the average annual releases, over a 10-year period, this amounts to more than 1 million lbs. of PM_{10} releases as fugitive particles. Releases of such magnitude cannot be ignored. Sampling conducted on properties off-site from the TVA plant showing the presence of fly ash particles are unto themselves not sufficient to link impacts in the community to the 2008 spill event.

The National Emissions Inventory provides particle size classes for the fugitive PM as reported in Table 5.3. The data reported in the NEI show that about 92% of the PM_{10} released as fugitive emissions are filterable particles, and only about 8% are condensible particles that are less than 1 micron in size. About 31% of the filterable mass is $PM_{2.5}$ (2.5 microns and smaller).

The Toxics Release Inventory (TRI) is another source of emissions reported by the facility to the U.S. EPA. The TRI is intended to keep communities informed of the release of hazardous pollutants from facilities. The TRI requires facilities to report estimated releases of regulated hazardous air pollutants (HAPs). There are approximately 650 HAPs that fall under the reporting requirements of the TRI, and among these, reporting is required only when a facility exceeds a certain threshold value. Unlike the NEI, the TRI reports all media releases and transfers, not just to air.

TRI reporting began in 1987; however, public utilities were exempt from meeting the reporting requirements until 1998. Table 5.4 tabulates the reported air emissions for the facility obtained from EPA's ToxNet website for the facility's reporting.

According to the NEI, about 5% of the total emissions from the facility are fugitive emissions, and all of these are in the form of PM. According to the TRI, which is what communities generally rely on to be kept informed of HAPs, TVA reports fugitive emissions to be almost insignificant (on the average about 0.12%). The facility also reports small but significant releases of dioxins. Dioxins condense to form particulate matter and hence they are a part of the composition of fly ash.

For other pollution media, the facility reports off-site transfers, transfers to on-site landfills, transfers to surface impoundments, and releases to surface water bodies (Emory River). For its 2008 reporting year, it reported the spill as

Table 5.3 PM Size Classes Derived from the NEI Reported Data

PM Size Class	Mass (Lbs.)	Wt. Fraction
Fugitive PM_{10} (Filterables + Condensibles)	107,329	1.00
Primary PM_{10}, Filterable Portion Only	98,912	0.92
Primary $PM_{2.5}$ (Includes Filterables + Condensibles)	38,703	0.36
Primary $PM_{2.5}$, Filterable Portion Only	30,286	0.31 of filterable mass
Primary PM Condensible Portion Only (All < 1 Micron)	8,418	0.08

Table 5.4 TRI Reported Releases to Air

Chemical Name	Unit	2009	2008	2007	2006	2005	2004	2003	2002	2001	2000	1999	1998	Total
Fugitive Releases to Air														
1,2,4-Trimethylbenzene	Pounds	-	-	-	-	-	-	-	250	250	250	250	250	1,250
Ammonia	Pounds	2,600	11,000	7,000	12,000	11,000	1,300	-	5	-	-	-	-	44,905
Arsenic	Pounds	-	5	5	5	5	5	5	5	5	5	5	5	55
Barium	Pounds	7	250	250	250	250	250	250	250	250	250	250	250	2,757
Benzo(G,H,I) Perylene	Pounds	-	-	-	-	0	-	-	-	-	-	-	-	0
Chromium	Pounds	-	250	5	5	5	250	250	250	5	5	5	5	1,035
Cobalt	Pounds	-	5	5	5	5	5	5	5	5	5	5	5	55
Copper	Pounds	1	5	5	5	5	5	5	5	5	5	5	5	56
Hydrochloric Acid	Pounds	-	-	-	-	5	5	5	5	5	5	5	5	40
Hydrogen Fluoride	Pounds	-	-	-	-	5	5	5	5	5	5	5	5	40
Lead	Pounds	3	5	2	2	3	6	4	4	2	5	5	5	46
Manganese	Pounds	71	250	250	250	250	250	250	250	250	250	250	250	2,821
Mercury	Pounds	-	-	-	-	0	0	0	0	-	0	-	-	1
N-Hexane	Pounds	-	-	-	-	-	-	-	250	250	250	250	250	1,250
Nickel	Pounds	-	250	5	5	5	250	5	5	5	5	5	5	545
PACs*	Pounds	0.1	0.1	0.1	0.1	0.77	0.8	0.8	0.8	0.4	0.8	0	0	4.77
Selenium	Pounds	-	-	-	-	-	5	5	5	5	5	5	5	35
Styrene	Pounds	-	5,400	-	-	-	-	-	-	-	-	-	-	5,400
Sulfuric Acid	Pounds	-	-	-	-	5	5	5	5	5	5	5	250	285
Vanadium	Pounds	2	5	5	5	5	5	5	5	5	5	-	-	47
Zinc	Pounds	1	5	5	5	5	5	5	5	5	5	5	5	56
Total Air Fugitives	Pounds	2,685.1	17,429.6	7,537.3	12,537	11,554	2,352	805	1,309.8	1,057.8	1,060.9	1,055.0	1,300.0	60,683.7
Stack Releases To Air														
1,2,4-Trimethylbenzene	Pounds	0	0	0	0	0	0	0	250	250	250	250	250	1250
Ammonia	Pounds	4,800	27,000	14,000	23,000	18,000	-	-	-	-	-	-	-	86,800
Arsenic	Pounds	-	250	250	250	250	250	250	250	1,500	2,000	2,100	2,100	9,450
Barium	Pounds	460	2,900	2,400	1,600	1,500	750	750	750	1,000	1,800	1,800	1,900	17,610
Benzo(G,H,I)Perylene	Pounds	-	0	0	0	0	0	0	0	0	0	-	-	1

(*Continued*)

Table 5.4 TRI Reported Releases to Air—*cont'd*

Chemical Name	Unit	2009	2008	2007	2006	2005	2004	2003	2002	2001	2000	1999	1998	Total
Chromium	Pounds	-	250	250	250	250	250	750	750	750	250	750	750	5,250
Cobalt	Pounds	-	250	250	250	250	250	250	250	250	250	250	250	2,750
Copper	Pounds	89	250	250	250	250	250	250	250	750	750	750	750	4,839
Hydrochloric Acid	Tons	385	1,550	1,700	1,850	1,650	2,100	2,250	2,300	2,000	2,100	2,150	3,050	23,085
Hydrogen Fluoride	Tons	48	225	225	250	225	245	250	265	255	270	275	300	2,833
Lead	Pounds	61	280	290	240	220	310	330	380	240	250	250	250	3,101
Manganese	Pounds	140	750	750	750	750	750	750	750	750	250	250	250	6,890
Mercury	Pounds	100	480	480	480	430	450	490	480	450	510	-	-	4,350
N-Hexane	Pounds	-	-	-	-	-	-	-	250	250	250	250	250	1,250
Naphthalene	Pounds	13	250	250	250	250	250	-	-	-	-	-	-	1,263
Nickel	Pounds	-	250	250	250	250	250	750	750	250	250	250	250	3,750
PACs*	Pounds	7	32	32	32	29	27	28	30	27	29	-	-	273
Selenium	Pounds	-	-	9,000	10,000	-	11,000	12,000	12,000	7,700	9,100	9,200	9,400	89,400
Styrene	Pounds	-	250	-	-	-	-	-	-	-	-	-	-	250
Sulfuric Acid	Tons	100	285	330	340	390	500	425	1,100	700	420	500	550	5,640
Vanadium	Pounds	92	250	250	250	250	250	750	750	750	750	-	-	4,342
Zinc	Pounds	320	1,500	1,500	1,400	1,300	1,500	1,600	1,700	250	250	250	250	11,820
Total Air Stack (Excluding Dioxins)	Tons	536.0	2,077.5	2,270.1	2,459.6	2,277.0	2,853.3	2,934.5	3,674.8	2,962.6	2,798.5	2,933.2	3,908.3	31,685.3
Total Air (Stack + Fugitive; Excluding Dioxins)	Tons	537.4	2,086.2	2,273.9	2,465.9	2,282.8	2,854.4	2,934.9	3,675.4	2,963.1	2,799.0	2,933.7	3,909.0	31,715.7
Percent of Releases as Fugitive	Percent	0.25	0.42	0.17	0.25	0.25	0.04	0.01	0.02	0.02	0.02	0.02	0.02	0.10
Dioxin—Fugitive Releases	Grams	NR	NR	NR	NR	0.18	0.18	0.18	0.18	0.08	0.08	NR	NR	0.98
Dioxin—Stack Releases	Grams	0.45	2.24	2.20	2.20	2.00	1.9	2.0	2.05	1.9	3.9	NR	NR	20.8

*PACs—Polycyclic Aromatic Compounds
Source: ToxNet

a release to the surface water body, and no releases to air. Table 5.5 tabulates TVA's reported releases to the surface water body beginning in 1998 through 2009.

The last column in Table 5.5 provides a comparison of the releases reported from the 2008 spill to the total releases reported over the previous 10-year period. Overall, the December 2008 release was equivalent to 4.5 times more toxins reported as being released than the cumulative discharges between 1998 and 2007.

5.8 Airborne Releases from the Spill Incident

According to TVA's reporting under the TRI, there have been no releases of toxic chemicals into the community via the air pathway. TVA has reported that 100% of the ash spilled from the collapsed surface impoundment was emptied into the Emory River.

Throughout the post-spill incident, TVA made air quality measurements both on its property, in areas where it was performing dredging and excavating operations, and in various locations within the immediate community in general proximity to the spill site. There are nearly 268,000 data entries for air measurements reported by a handful of workers that recorded PM_{10} at different locations. The exact locations where the measurements were obtained are not reported except by a coordinate system that is not apparent from the database and two monitors that were fixed. Measurements were usually taken at 1-hour intervals, but sometimes over longer and shorter intervals. Although the exact coordinates of the measurements were not reported in the TVA database, the measurements indicate general areas such as residential sites, at churches and schools, the community center, various businesses, the park, and along certain roadsides. Included in the measurements are anecdotal statements describing environmental conditions at the time a measurement was taken, such as odors, visible smoke, winds, fog, mist, rain, presence of trucks, and other descriptors. In a number of instances, visible dust was reported along with the measurements; however, some of the statements are accompanied by additional statements that link the airborne dust to the neighboring quarry operations despite the fact that no analytical sampling was performed to confirm whether the dust was quarry dust or coal combustion products.

In order to assess whether the community has been impacted by fugitive dust resulting from coal combustion products, the database was sorted to distinguish non-coal combustions product dust, as well as conditions that might impact on the accuracy of the PM measurements. Brush fires, wood burning, barbeques, or cigarette smoke in the proximity of monitors used for

Table 5.5 TRI Reported Releases to Water in Pounds

Chemical Name	2009	2008	2007	2006	2005	2004	2003	2002	2001	2000	1999	1998	Total Reported 1998–2007	Ratio 1998–2007 Releases to 2008
Arsenic	NR	142,600	2,700	11,000	9,600	9,200	8,900	9,000	11,000	3,500	4,300	4,800	74,000	1.9
Barium	38,000	692,000	44,000	50,000	44,000	42,000	40,000	41,000	15,000	13,000	16,000	18,000	323,000	2.1
Chromium	NR	64,500	1,600	750	750	750	750	750	250	1,800	2,200	2,400	12,000	5.4
Cobalt	NR	46,000	—	—	—	NR	NR	NR	250	1,800	2,200	2,400	6,650	6.9
Copper	4,000	98,000	4,000	3,700	3,700	3,700	3,700	3,700	2,400	2,800	2,900	750	31,350	3.1
Lead	—	71,000	—	—	—	NR	NR	NR	34	1,800	2,200	2,400	6,434	11.0
Manganese	—	750,000	—	—	—	NR	NR	NR	750	9,200	11,000	13,000	33,950	22.1
Mercury	—	400	—	—	—	NR	NR	NR	7	7	NR	NR	14	29.0
Nickel	NR	73,250	250	—	—	NR	NR	NR	250	1,800	2,200	2,400	6,900	10.6
Selenium	NR	NR	1,100	3,000	NR	2,500	2,400	2,400	1,500	3,500	4,300	4,800	25,500	NR
Vanadium	3,200	643,200	3,400	8,200	7,100	6,500	6,300	7,600	14,000	15,000	NR	NR	68,100	9.4
Zinc	3,700	184,000	4,000	1,800	1,700	1,700	1,700	1,700	1,200	3,800	4,400	3,800	25,800	7.1
Totals	48,900	2,764,950	61,050	78,450	66,850	66,350	63,750	66,150	46,641	58,007	51,700	54,750	613,698	4.5

PM measurements result in high readings that would confound any attempts to determine whether the ash from the spill had become airborne. Other contributors to fugitive dust include traffic from heavy machinery, roadwork, construction activities, and other activities. In order to compare on-site to off-site measurements of PM to assess whether a correlation exists between the area where the ash spilled and the surrounding community, the database was first sorted to separate out the effects of non-coal combustion sources of airborne dust. This sorting was done for each general location where measurements were obtained.

Figure 5.10 shows the effects of different sources and conditions on PM concentration for one data set considered in the analysis. This bar chart compares the mean PM_{10} measurements taken between December 2008 and July 2010 within the community for different conditions noted by the observer. The bar chart is illustrative only, showing that there are significant differences between the data set populations.

While the bar chart is qualitatively suggestive of how the order of PM concentrations should follow, this is not always the case. For example, the presence of smoke was found to be considerably greater than all other

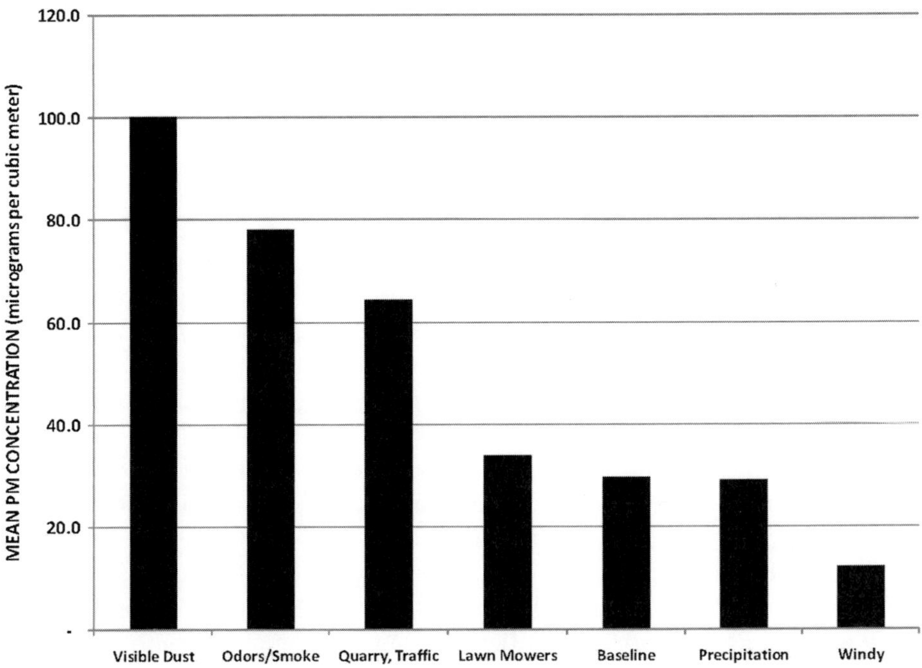

Figure 5.10 The effect of different sources and local conditions on PM measurements for measurements obtained near businesses.

measurements, whereas quarry dust and the presence of lawn mower operations were found in some instances to be least significant.

Table 5.6 reports the statistical parameters of the entire database examined in which the subset measurements for each location have been sorted from the highest mean to lowest mean value. Comparisons between data sets for different locations and even within data sets for any given location show wide variability as measured by the range of PM measurements (the difference between the highest and lowest measurement).

While the large variability suggests that the measurements are somewhat random in nature and highly influenced by different local conditions, a close examination of the statistical values shows in fact that this is not the case. Further, the table reports the frequency of different observed conditions within each data subset. Events such as fire, odors, visible smoke, high traffic conditions, construction, and other observed conditions and confounders represent infrequent events and hence do not have a significant effect on drawing conclusions when making comparisons between the measurements made at different locations.

The line chart in Figure 5.11 compares the daily average PM_{10} measurements reported from within the TVA property to those measured at different residences. The time period of the measurements covers from December 28, 2008, through September 27, 2009. The PM data considered in the analysis excludes the confounders such as odors, smoke, fires, windy conditions, precipitation, and dust linked to activities such as the quarry and heavy traffic or construction. The chart is based on comparing baseline air quality conditions between the two general site locations.

Since the data are almost indistinguishable, the plot shows that both on-site and off-site PM_{10} measurements are strongly related. While Figure 5.11 shows that the PM measurements made on-site versus off-site correlate with each other, the plot does not show whether one site influences the other.

To determine the influence of one site upon the other, I examined a timeline that TVA prepared showing the level of activities it performed on-site between March 26, 2009, and September 27, 2009 (see Figure 5.12). From this chart information on the amounts of ash excavated and dredged, I extracted notable site conditions and then overlaid these observations on the plot depicted in Figure 5.11. This plot is shown in Figure 5.13.

Figure 5.13 shows a strong correlation between the on-site dredging and excavation activities, as well as impacts of idle periods and flooding with the daily average PM_{10} levels within the community. The chart confirms for me that the TVA site conditions and activities resulted in the release of PM into the community.

Table 5.6 Statistical Parameters Characterizing PM_{10} Measurements Made by TVA

Set No.	Description	1st Meas.	Last Meas.	No. Meas.	Frequency Within Set	Min PM (μg/m³)	Max PM (μg/m³)	Mean PM (μg/m³)	Range (μg/m³)
1	Community—evacuation	9/17/2009	7/17/2010	9	28.1	5.0	72.0	26.9	67.0
2	TVA Site—areas evacuated	3/13/2009	7/16/2010	23	71.9	8.0	65.0	27.4	57.0
9	Business—visible dust	2/17/2009	1/14/2010	17	0.1	23.0	330.0	100.1	307.0
4	Business—odors, smoke, fires	2/8/2009	7/14/2010	294	1.3	10.0	668.0	78.2	658.0
5	Business—quarry, trucks, traffics	9/4/2009	6/8/2010	28	0.1	10.0	166.0	64.5	156.0
6	Business—Mowers in area	9/5/2009	10/7/2009	5	0.0	11.0	42.0	34.0	31.0
8	Business—baseline	2/6/2009	7/23/2010	21915	97.0	—	240.0	29.7	240.0
3	Business and Community—precipitation	9/5/2009	7/17/2010	281	1.2	1.0	93.0	29.1	92.0
7	Business—winds	9/17/2009	4/13/2010	47	0.2	1.0	49.0	12.2	48.0
15	Church—visible dust	2/18/2009	6/6/2009	3	0.1	73.0	154.0	116.7	81.0
16	Church- odors, smoke, fires	2/17/2009	9/17/2009	17	0.8	14.0	164.0	52.0	150.0
13	Church—mowers	9/4/2009	10/22/ 2009	5	0.2	6.0	59.0	37.0	53.0
10	Church—baseline	2/8/2009	7/23/2010	1955	95.3	—	129.0	26.6	129.0
11	Church—precipitation	9/6/2009	7/17/2010	57	2.8	1.0	67.0	24.5	66.0
12	Church—Traffic	9/18/2009	10/6/2009	4	0.2	16.0	39.0	24.3	23.0
14	Church—winds	9/22/2009	11/4/2009	10	0.5	2.0	31.0	11.6	29.0
19	Community Center—smoke	7/4/2009	1/15/2010	3	1.3	39.0	138.0	78.7	99.0
17	Community Center—baseline	3/31/2009	7/20/2010	224	97.4	1.0	92.0	33.1	91.0
18	Community Center—precipitation and wind	12/14/ 2009	1/24/2010	3	1.3	4.0	50.0	28.3	46.0
20	Daycare—precipitation	9/20/2009	12/18/ 2009	5	5.1	6.0	59.0	31.4	53.0
23	Daycare—mowers	9/7/2009	9/7/2009	1	1.0	30.0	30.0	30.0	—
21	Daycare—baseline	5/10/2009	3/19/2010	91	91.9	2.0	56.0	19.6	54.0

(*Continued*)

Table 5.6 Statistical Parameters Characterizing PM_{10} Measurements Made by TVA—*cont'd*

Set No.	Description	1st Meas.	Last Meas.	No. Meas.	Frequency Within Set	Min PM ($\mu g/m^3$)	Max PM ($\mu g/m^3$)	Mean PM ($\mu g/m^3$)	Range ($\mu g/m^3$)
22	Daycare—wind	10/17/2009	10/25/2009	2	2.0	3.0	21.0	12.0	18.0
27	Hospital—gravel road	9/11/2009	9/11/2009	1	0.3	41.0	41.0	41.0	—
26	Hospital—baseline	4/1/2009	7/23/2010	356	97.3	1.0	97.0	27.5	96.0
24	Hospital—precipitation	9/8/2009	1/24/2010	7	1.9	1.0	32.0	15.1	31.0
28	Hospital—tank release	10/13/2009	10/13/2009	1	0.3	13.0	13.0	13.0	—
25	Hospital—wind	10/24/2009	10/24/2009	1	0.3	5.0	5.0	5.0	—
33	Park—odors, smoke, fire	4/4/2009	1/15/2010	7	0.4	25.0	210.0	86.0	185.0
34	Park—airborne dust	6/1/2009	6/1/2009	1	0.1	62.0	62.0	62.0	—
30	Park –mowers	9/8/2009	9/23/2009	2	0.1	17.0	69.0	43.0	52.0
32	Park—baseline	4/1/2009	7/23/2010	1509	95.1	1.0	95.0	26.2	94.0
29	Park—precipitation	9/8/2009	1/24/2010	54	3.4	2.0	58.0	21.8	56.0
31	Park—wind	9/25/2009	1/24/2010	14	0.9	1.0	40.0	14.1	39.0
38	TVA—fires, odors, smoke	1/4/2009	5/11/2010	329	0.4	2.0	5,060.0	86.9	5,058.0
35	TVA—visible dust	1/12/2009	8/1/2009	89	0.1	1.0	146.0	66.2	145.0
37	TVA—quarry, heavy traffic	9/8/2009	4/14/2010	8	0.01	19.0	165.0	56.5	146.0
39	TVA—precipitation	5/4/2009	7/17/2010	318	0.4	—	420.0	53.7	420.0
40	TVA—mowers	9/18/2009	6/8/2010	3	0.00	17.0	52.0	30.7	35.0
41	TVA—baseline	12/28/2008	7/23/2010	80161	99.0	—	275.0	26.9	275.0
36	TVA—winds	9/18/2009	4/16/2010	24	0.03	1.0	50.0	14.3	49.0
42	Residential—fires, odors	12/28/2008	7/22/2010	1159	0.8	1.0	1,800.0	85.8	1,799.0
43	Residential—quarry, high traffic	9/5/2009	7/3/2010	33	0.0	7.0	294.0	81.0	287.0

44	Residential—winds	1/29/2009	7/4/2009	112	0.1	14.0	207.0	76.8	193.0
45	Residence—baseline	12/28/ 2008	7/23/2010	149260	97.5	—	410.0	27.5	410.0
46	Residence—precipitation	5/4/2009	7/18/2010	2119	1.4	1.0	200.0	25.2	199.0
47	Residential—mowers	9/3/2009	4/16/2010	62	0.0	2.0	72.0	23.5	70.0
48	Residential—winds	9/9/2009	4/16/2010	388	0.3	—	62.0	9.9	62.0
55	Schools—visible dust	4/7/2009	7/30/2009	12	0.2	6.0	93.0	68.3	87.0
51	Schools—odors, fires	2/9/2009	7/15/2010	53	0.8	12.0	213.0	58.7	201.0
56	Schools—construction	6/29/2010	9/12/2009	2	0.0	46.0	62.0	54.0	16.0
52	Schools—quarry, trucks	6/15/2010	9/17/2009	2	0.0	28.0	79.0	53.5	51.0
49	School—baseline	2/8/2009	7/23/2010	6242	96.9	—	139.0	26.4	139.0
53	Schools—mowers	9/9/2009	7/21/2010	9	0.1	9.0	40.0	22.2	31.0
50	School—precipitation	9/3/2009	7/22/2010	104	1.6	—	67.0	20.8	67.0
54	Schools—winds	9/16/2009	7/21/2010	17	0.3	4.0	34.0	17.1	30.0

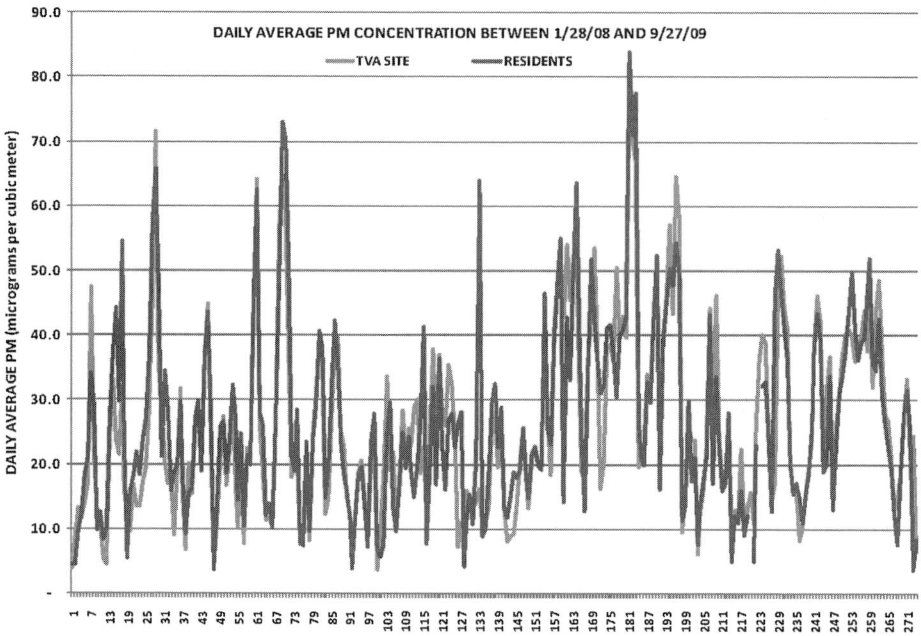

Figure 5.11 Comparison of on-site and off-site PM_{10} measurements.

Figures 5.14 and 5.15 compare the baseline monthly average PM_{10} measurements regressed from TVA's data for different sampling area locations. These comparisons provide further confirmation that the off-site particulate measurements correlate with measurements reported from within the property line of the TVA facility.

5.9 Exceedances and Excursions

The short-term (24-hour or daily average) National Ambient Air Quality Standard (NAAQS) for PM_{10} is 150 micrograms per cubic meter of air ($\mu g/m^3$). The NAAQS for $PM_{2.5}$ is 35 $\mu g/m^3$. There is no evidence in the database reviewed to support that the PM_{10} NAAQS had been exceeded.

Several other standards were considered. The Occupational Safety and Health Administration (OSHA) permissible exposure limits (PELs) for "mineral dusts" listed here are from Table Z-3 of 29 CFR 1910.1000. The OSHA PEL (8-hour TWA) for crystalline silica (as respirable quartz) is either 250 mppcf divided by the value "$\%SiO_2 + 5$" or 10 mg/m^3 divided by the value "$\%SiO_2 + 2$." The OSHA PEL (8-hour TWA) for crystalline silica (as total quartz) is 30 mg/m^3 divided by the value "$\%SiO_2 + 2$." The OSHA PELs

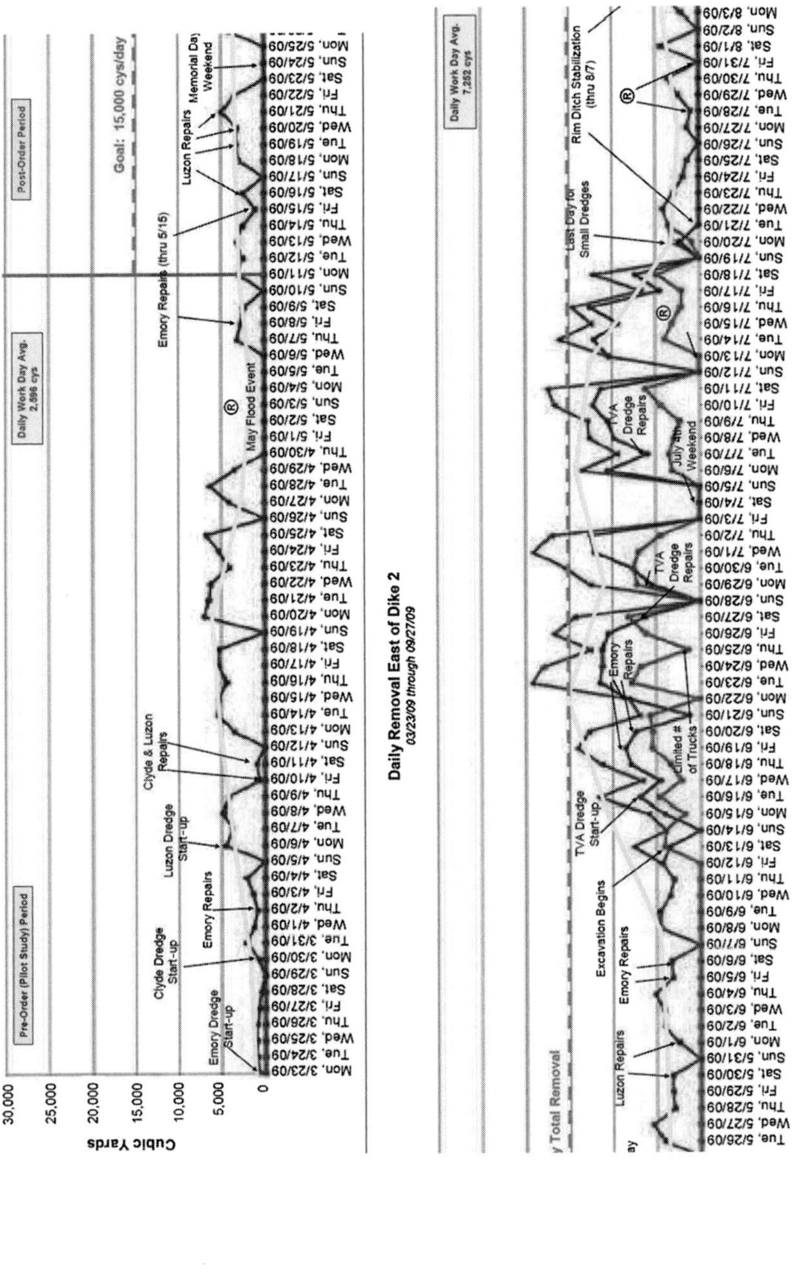

Figure 5.12 TVA timeline of activities between March 26 and September 27, 2009.

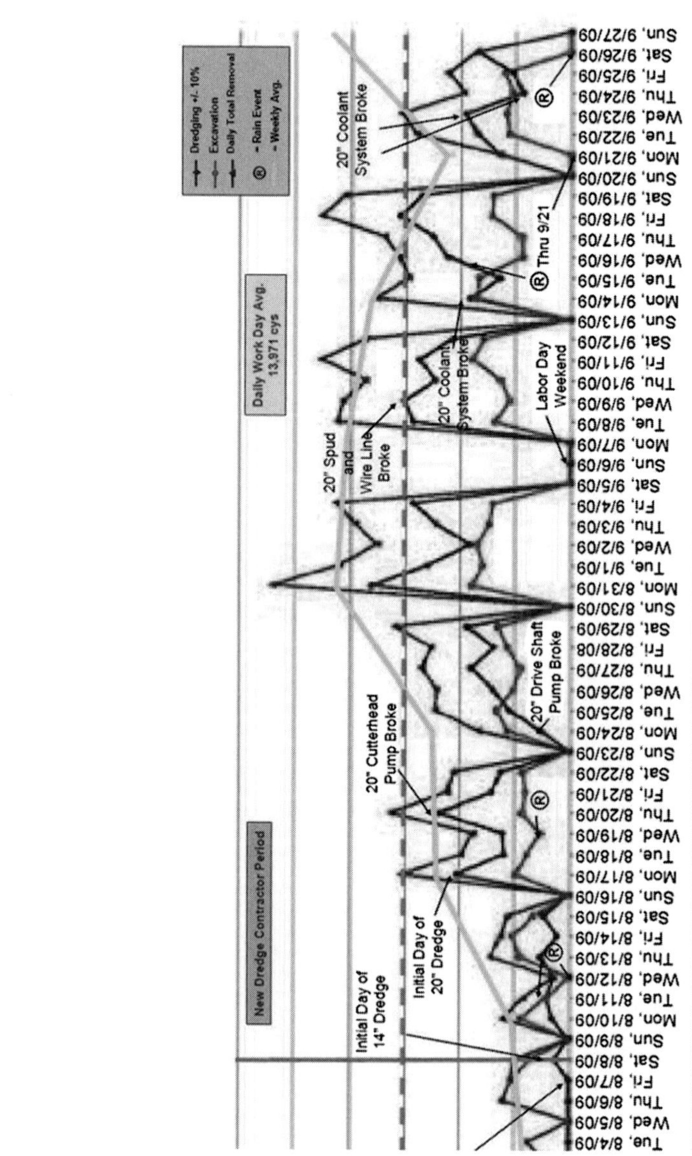

Figure 5.12 *(continued).*

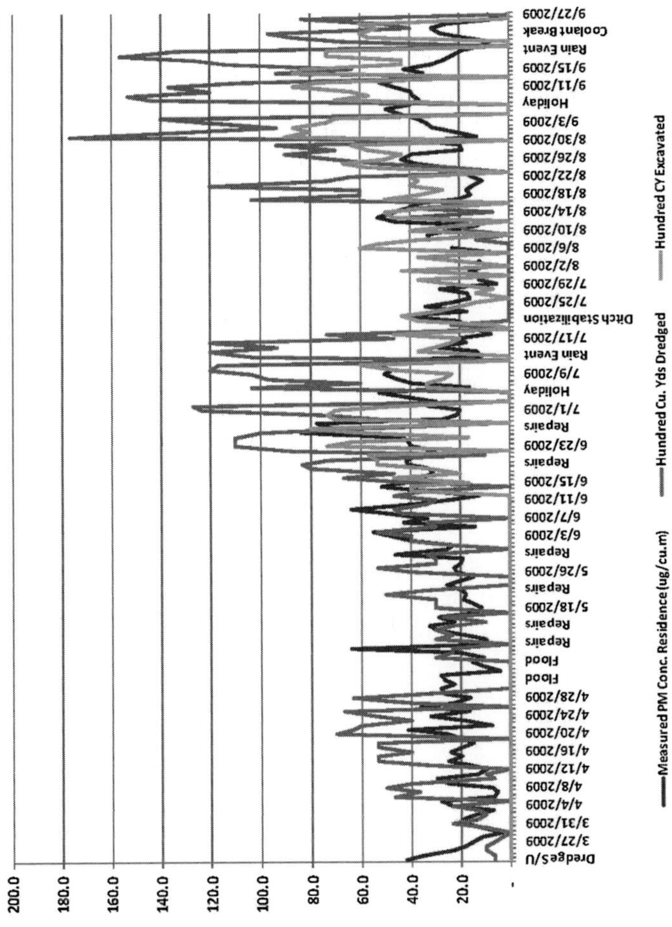

Figure 5.13 Plot showing site activities influencing PM concentrations at residential locations.

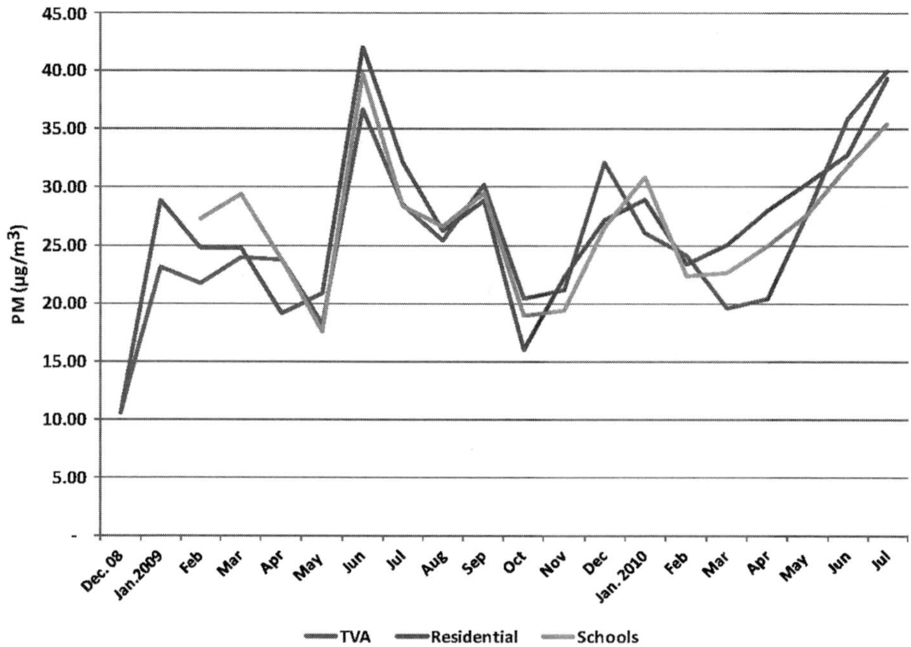

Figure 5.14 Comparison between monthly average PM_{10} measurements for schools and residences to site measurements. Confounders are omitted.

(8-hour TWAs) for cristobalite and tridymite are one-half the values calculated above using the count or mass formulae for quartz. The OSHA PEL (8-hour TWA) for amorphous silica (including diatomaceous earth) is either 80 mg/m^3 divided by the value "%SiO_2," or 20 mppcf. The OSHA PELs (8-hour TWAs) for talc (not containing asbestos), mica, and soapstone are 20 mppcf. The OSHA PEL (8-hour TWA) for Portland cement is 50 mppcf. The OSHA PEL (8-hour TWA) for graphite (natural) is 15 mppcf. The PELs for talc (not containing asbestos), mica, soapstone, and Portland cement are applicable if the material contains less than 1% crystalline silica. The OSHA PEL (8-hour TWA) for coal dust (as the respirable fraction) containing less than 5% SiO_2 is 2.4 mg/m^3 divided by the value "%SiO_2 + 2." The OSHA PEL (8-hour TWA) for coal dust (as the respirable fraction) containing greater than or equal to 5% SiO_2 is 10 mg/m^3 divided by the value "%SiO_2 + 2." Again, I found no evidence to support that these standards were exceeded.

The database of PM_{10} values showed some but not large numbers of excursions above the safe levels of exposures. Higher frequencies of excursions of PM_{10} were observed from Dust Track monitoring that was conducted by the author's staff. The TVA database indicates that the particulate levels in

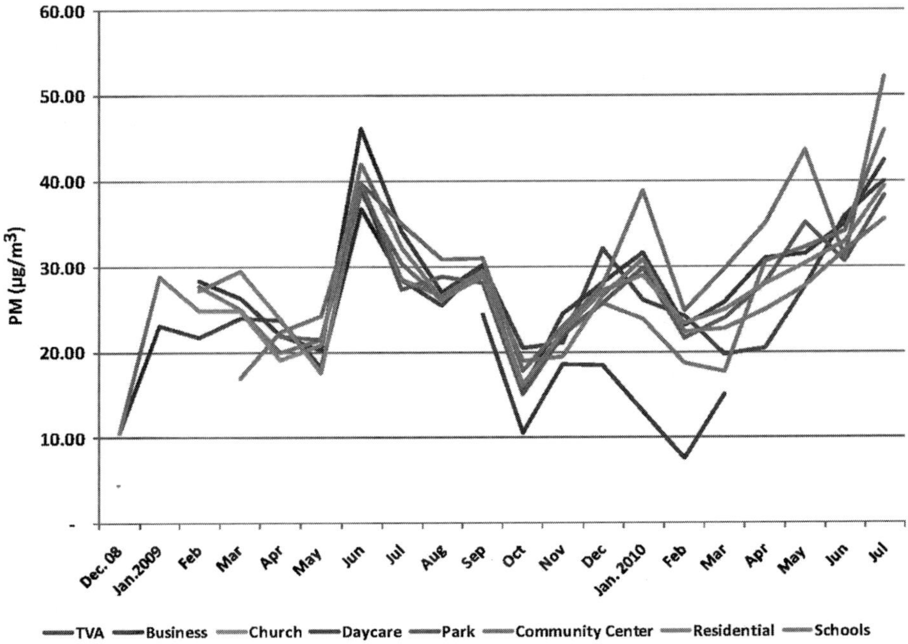

Figure 5.15 Comparison between monthly average PM_{10} measurements for different off-site locations to site measurements. Confounders are omitted.

the area around the TVA ash release remained below EPA standards for ambient air quality throughout the monitored periods of the response. Rain and moisture in the ash are factors that contributed to the low readings.

The database shows that for the majority of measurements reported, World Health Organization air quality guidelines were exceeded. A threshold value of 20 $\mu g/m^3$ is considered by the WHO to be the lowest level at which total, cardiovascular, and lung cancer have been shown to increase with more than 95% confidence in $PM_{2.5}$.[16]

5.10 Toxins Released

Although PM values did not exceed NAAQS, nonetheless, fly ash was released into the community based on the observations reported previously. Fly ash contains a broad array of toxins including but not limited to arsenic, barium, beryllium, boron, cadmium, chromium, chromium VI, cobalt, copper, fluorine,

[16] WHO Air Quality Guidelines: Global Update 2005: Particulate Matter, Ozone, Nitrogen Dioxide and Sulfur Dioxide. 2006. <http://www.euro.who.int/data/assets/pdf_file/0005/78638/E90038.pdf>

lead, manganese, nickel, selenium, strontium, thallium, vanadium, zinc, and other hazardous constituents. Based on the literature of reported compositions tabulated in Table 5.2, the average composition of each of these constituents was calculated using the mean and maximum PM_{10} concentrations for the baselines recorded for each general site location. These concentrations were then compared against the safe air quality guidelines recommended by the World Health Organization. The analysis is summarized in Table 5.7.

Table 5.7 tabulates the calculated mean and maximum concentration of toxins for different locations and compares these values against the WHO's 2005 guidelines. Highlighted values indicate exceedances and/or excursions which support that the guidelines were exceeded at times. The analysis shows that the community was exposed to arsenic, cadmium, chromium (total and valence VI), lead, manganese, nickel, and vanadium at levels that either exceeded the WHO guidelines (in some instances) or were above national averages for rural areas.

It is important to note that the analysis does not necessarily support an elevated health risk to the community—only that safe air quality guidelines were exceeded and/or elevated over norms. Many of the exceedances noted in this analysis are based on maximum PM_{10} concentrations reported by TVA's contractors. The empirical data show these to be low-frequency events over the measurement periods.

5.11 An Estimate of the Releases to Air

An estimate of the amount of ash released into the air from the spill was made from an analysis of regional facility reporting information on PM. Table 5.8 tabulates the reported TRI releases to air reported by all of the industrial facilities in the region. The TRI data for air emissions are calculated by facilities for point (stack) and fugitive source releases.

Table 5.8 shows that the Kingston Fossil Plant (KIF) is the largest polluter in the region (an area that is within a 40-mile radius of the KIF plant) and that combined TVA's two coal-burning plants account for more than 56% of the air emissions to the region. Consequently, they directly influence the air quality of the area. To better illustrate this, Table 5.9 tabulates percentage of air releases reported by all facilities in the region. The same table also shows the change in reported releases between 2008 and 2009.

Figure 5.16 shows a plot of TRI reported releases for all facilities between 2004 and 2009. This figure shows that TVA's fossil-burning plants account for the majority of air pollution in the region and further that both plants as well as most of the industrial facilities in the region reported significant

Table 5.7A Calculated Toxin Concentrations in Air*

	Concentration in ng/m³															
	Business Areas		Churches		Community Center		Daycare Center		Hospital		Park		Residences		School	
Toxin	Mean	Max	Mean	Max	Mean	Max	Mean	Max	Mean	Max	Mean	Max	Mean	Max	Mean	Max
Arsenic	2.2	18.0	2.0	9.7	2.5	6.9	1.5	4.2	2.1	7.3	2.0	7.1	2.1	30.8	2.0	10.4
Barium	23.9	193.4	21.4	104.0	26.6	74.2	15.8	45.1	22.1	78.2	21.1	76.6	22.2	330.5	21.3	112.0
Beryllium	0.1	1.2	0.1	0.6	0.2	0.5	0.1	0.3	0.1	0.5	0.1	0.5	0.1	2.1	0.1	0.7
Boron	18.7	150.7	16.7	81.0	20.8	57.8	12.3	35.2	17.2	60.9	16.5	59.7	17.3	257.5	16.6	87.3
Cadmium	1.9	15.6	1.7	8.4	2.2	6.0	1.3	3.6	1.8	6.3	1.7	6.2	1.8	26.7	1.7	9.0
Chromium	15.0	121.2	13.4	65.1	16.7	46.5	9.9	28.3	13.9	49.0	13.2	48.0	13.9	207.1	13.4	70.2
Chromium VI	2.7	21.6	2.4	11.6	3.0	8.3	1.8	5.0	2.5	8.7	2.4	8.6	2.5	36.9	2.4	12.5
Cobalt	7.8	63.2	7.0	34.0	8.7	24.2	5.2	14.8	7.2	25.6	6.9	25.0	7.2	108.0	7.0	36.6
Copper	3.3	26.9	3.0	14.4	3.7	10.3	2.2	6.3	3.1	10.9	2.9	10.6	3.1	45.9	3.0	15.6
Fluorine	0.9	7.0	0.8	3.7	1.0	2.7	0.6	1.6	0.8	2.8	0.8	2.8	0.8	11.9	0.8	4.0
Lead	74.3	600.4	66.6	322.7	82.7	230.1	49.1	140.1	68.7	242.7	65.5	237.6	68.8	1,025.6	66.2	347.7
Manganese	45.4	367.0	40.7	197.2	50.5	140.7	30.0	85.6	42.0	148.3	40.1	145.3	42.0	626.9	40.4	212.5
Mercury	4.5E-03	3.6E-02	4.0E-03	1.9E-02	5.0E-03	1.4E-02	2.9E-03	8.4E-03	4.1E-03	1.5E-02	3.9E-03	1.4E-02	4.1E-03	6.2E-02	4.0E-03	2.1E-02
Nickel	64.0	516.7	57.3	277.7	71.2	198.1	42.2	120.6	59.1	208.8	56.4	204.5	59.2	882.8	56.9	299.3
Selenium	1.3	10.7	1.2	5.8	1.5	4.1	0.9	2.5	1.2	4.3	1.2	4.2	1.2	18.3	1.2	6.2
Strontium	23.0	186.0	20.6	100.0	25.6	71.3	15.2	43.4	21.3	75.2	20.3	73.6	21.3	317.8	20.5	107.7
Thallium	0.3	2.2	0.2	1.2	0.3	0.8	0.2	0.5	0.2	0.9	0.2	0.9	0.2	3.7	0.2	1.3
Vanadium	7.5	60.5	6.7	32.5	8.3	23.2	4.9	14.1	6.9	24.4	6.6	23.9	6.9	103.3	6.7	35.0
Zinc	52.1	421.2	46.7	226.4	58.0	161.5	34.4	98.3	48.2	170.2	46.0	166.7	48.2	719.6	46.4	243.9

*Highlighted values exceed WHO guidelines or are elevated values. See comments and WHO guidelines in Table 5.7B.

Table 5.7B Conclusions Based on Comparisons Made Between Calculated Toxin Concentrations and Air Quality Guidelines

Toxin	Overall Mean (ng/m^3)	Overall Max (ng/m^3)	Guidelines	Conclusions
Arsenic	2.0	11.8	WHO (2005)—Airborne concentrations of arsenic range from 1 ng/m^3 to 10 ng/m^3 in rural areas and from a few nanograms per cubic meter to about 30 ng/m^3 in noncontaminated urban areas.	Values generally fall within typical ranges. Excursion supports slight to moderately high elevation.
Barium	21.8	126.7	ACGIH TLV-TWA 0.5 mg/m^3; NIOSH REL (TWA) 0.5 mg/m^3; OSHA 8-hr TWA 0.5 mg/m^3	Not exceeded
Beryllium	0.1	0.8	OSHA 2 ug/m^3 8-hr TWA	Not exceeded
Boron	17.0	98.8	OSHA PEL 15 mg/m^3 total dust; 5 mg/m^3 respirable dust	Not exceeded
Cadmium	1.8	10.2	IARC Group 1. WHO Guideline 5 ng/m^3	Values generally fall within typical ranges. Excursion supports slight to moderately high elevation.
Chromium	13.7	79.4	WHO (2005)—IARC Group 1. Chromium is ubiquitous in nature. Available data, generally expressed as total chromium, show a concentration range of 5–200 ng/m^3. At an air concentration of chromium (VI) of 1 $\mu g/m^3$, the lifetime risk is estimated to be 4×10^{-2}.	Values exceeded supporting community risk.
Chromium VI	2.4	14.2	IARC Group 1. No safe level of chromium (VI) can be recommended.	Values exceeded supporting community risk.
Cobalt	7.1	41.4	OSHA PEL 0.1 mg/m^3	Not exceeded
Copper	3.0	17.6	OSHA PEL 1 mg/m^3	Not exceeded
Fluorine	0.8	4.6	OSHA TWA 0.2 mg/m^3	Not exceeded

Lead	67.7	393.4	WHO (2005)—Guidelines for lead in air will be based on the concentration of lead in blood.	Unknown but likely exceeded
Manganese	41.4	240.4	WHO (2005)—In urban and rural areas without significant manganese pollution, annual averages are mainly in the range of 10–70 ng/m^3; near foundries the level can rise to an annual average of 200–300 ng/m^3 and, where ferro- and silico-manganese industries are present, to more than 500 ng/m^3, with individual 24-hour concentrations sometimes exceeding 1,000 ng/m^3. WHO Guideline 150 ng/m^3.	Excursions exceed guidelines.
Mercury	4.1E-03	2.4E-02	Who (2005)—1 ug/m^3	Not exceeded
Nickel	58.3	338.6	WHO (2005)—Nickel levels in the ambient air are in the range 1–10 ng/m^3 in urban areas, although much higher levels (110–180 ng/m^3) have been recorded in heavily industrialized areas and larger cities. Assuming a linear dose–response, no safe level for nickel compounds can be recommended.	All levels deemed unsafe by WHO guidelines.
Selenium	1.2	7.0	OSHA PEL = 0.2 mg/m^3; ACGIH TLV = 0.2 mg/m^3	Not exceeded
Strontium	21.0	121.9	OSHA PEL = 15 mg/m^3 as total dust; 5 mg/m^3 respirable fraction for nuisance dust; ACGIH (TLV) 10 mg/m^3 as total dust	Not exceeded
Thallium	0.2	1.4	OSHA PEL 0.1 mg/m^3 (skin). Thallium is the most toxic of the heavy metals.	Not exceeded

(*Continued*)

Table 5.7B Conclusions Based on Comparisons Made Between Calculated Toxin Concentrations and Air Quality Guidelines—*cont'd*

Toxin	Overall Mean (ng/m^3)	Overall Max (ng/m^3)	Guidelines	Conclusions
Vanadium	6.8	39.6	WHO (2005)—Vanadium concentrations recorded in rural areas varied from a few nanograms to tenths of a nanogram per cubic meter, and in urban areas from 50 ng/m^3 to 200 ng/m^3. In cities during the winter, when fuel oil with a high vanadium content was used for heating, concentrations as high as 2000 ng/m^3 were reported. Air pollution by industrial plants may be less than that caused by power stations and heating equipment. WHO Guideline—Available data from occupational studies suggest that the LOAEL of vanadium can be assumed to be 20 $\mu g/m^3$, based on chronic upper respiratory tract symptoms. Since the adverse nature of the observed effects on the upper respiratory tract was minimal at this concentration, and a susceptible subpopulation has not been identified, a protection factor of 20 was selected. It is believed that below 1 $\mu g/m^3$ (averaging time 24 hours) environmental exposure to vanadium is not likely to have adverse effects on health.	Maximum values in data sets exceeded in a number of cases supporting high-level excursions and risks to community.
Zinc	47.5	276.0	OSHA (zinc chloride) 1 mg/m^3; PEL—TWA 5 mg/m^3; NIOSH 1 mg/m^3	Not exceeded

Ranking	Facility	TRI ID	Distance from Kingston, miles	Total Air Emissions in Pounds						
				2009	**2008**	**2007**	**2006**	**2005**	**2004**	**Total (2004–2009)**
1	US TVA Kingston Fossil Plant	37763STVKNSWANP	0	1,074,767	4,172,372	4,547,740	4,931,790	4,565,533	5,708,889	25,001,090
2	US TVA Bull Run Fossil Plant	37716STVBL1265E	22.77	237,782	2,796,224	4,598,427	3,336,102	4,557,277	3,368,289	18,894,100
3	Viskase Corp	37774VSKSCEASTL	15.7	1,922,124	1,897,168	1,874,173	1,778,178	2,116,148	2,258,147	11,845,938
4	ALCOA	37701LMNMCPOBOX	32.09	568,881	1,428,707	1,308,592	1,368,275	1,346,676	1,478,223	7,499,355
5	Sea Ray Boats Tellico Facility	37885SRYBT100SE	24.7	91,681	264,164	433,570	439,621	434,530	455,856	2,119,422
6	Sea Ray Boats Knoxville Facility	37914SRYBT2601S	38.86	52,424	180,572	260,652	323,333	416,203	444,087	1,677,271
7	Tate & Lyle Loudon	37774STLYMROUTE	16.41	224,069	262,715	262,018	267,333	360,567	207,715	1,584,417
8	Skier's Choice Inc	37801SCHLG1717H	30.17	57,436	154,249	160,246	376,200	353,000	290,000	1,391,131
9	Tennessee Watercraft Inc	37885CBBTC2000C	25.28	69,316	182,853	221,273	203,049	157,938	299,952	1,134,381
10	Mastercraft Boat Co	37885MSTRC100CH	24.73	49,207	218,944	269,400	206,949	203,048		947,548
11	NNSA Y-12 National Security Complex	37831SDKRDBEARC	16.78	141,354	82,012	76,810	192,024	87,759	230,825	810,783
12	American Safety Razor Co LLC	37914MRCNS2820M	39.84	49,378	114,089	84,526	148,971	164,698	129,632	691,294
13	Malibu Boats LLC	37774MLBBTHIGHL	14.7	58,019	100,429	141,120	115,827	112,029	153,522	680,946
14	Acupowder TN LLC	37742GRNBCBOX63	27.77	11,873	72,888	32,127	12,605	148,476	144,659	422,628
15	Rohm & Haas Chemicals LLC Knoxville Site	37921RHMND730DA	33.53	52,164	50,257	50,046	53,892	52,310	48,185	306,855
16	Johns Manville	37331MNVLLHWY41	36.19	11,655	13,183	9,601	8,671	48,083	193,156	284,349
17	Hubbell Lenoir City Inc	37771QZTCR300IN	16.82	37,575	43,262	47,167	43,860	48,779	42,977	263,620
18	Thyssenkrupp Waupaca Plant 6	37331THYSS134WA	36.85	29,519	46,640	43,540	44,640	44,740	39,540	248,619
19	Huber Engineered Woods LLC Whites Creek Facility	37381JMHBR1261D	17.37	4	29,757	58,701	47,929	51,432	60,664	248,487
20	Kimberly-Clark Corp Loudon Mill	37774KMBRL5600K	13.97	10,104	25,140	61,508	52,234	25,982	27,460	202,427
21	Ficosa NA	38555DLBRP87WYA	27.74		12,126	56,550	43,201	37,281	52,278	201,436

(Continued)

Table 5.8 TRI Reported Releases to Air for All Industry Within a 40-Mile Radius of the TVA Site—*cont'd*

Ranking	**Facility**	**TRI ID**	**Distance from Kingston, miles**	**Total Air Emissions in Pounds** 2009	**2008**	**2007**	**2006**	**2005**	**2004**	**Total (2004–2009)**
22	Denso Manufacturing Tennessee Inc	37801NPPND1720R	31.13	4,807	5,481	9,043	8,011	40,739	89,052	157,133
23	Omega Cabinetry	37716HMCRSHWY25	27.14		14,700	33,600	53,600	28,500	8,367	138,767
24	Vestal Manufacturing Enterprises Inc	37874VSTLM177IN	20.46	21,421	30,443	27,124	18,069	20,103	16,809	133,969
25	Toho Tenax America Inc	37854FRTFLROANE	6.78	5,396	26,097	37,088	30,726	28,914	—	128,221
26	Gerdau Ameristeel Knoxville	37921FLRDS1919T	32.8	12,193	12,537	12,192	11,669	11,134	44,563	104,288
27	Becromal of America	37716BCRML350JD	28.51	11,850	11,450	18,350	24,450	15,150	20,150	101,401
28	United States Utility Enclosures	37874NTDST223HW	23.74	23,556	23,263	35,935				82,754
29	La-Z-Boy Tennessee	37321LZBYTWALNU	36.66	15,647	13,797	10,279	12,980	14,560	15,162	82,424
30	Aleris Recycling Inc	37774MTLRSBLAIR	16.49	43,253	8,204	6,476	5,411	3,893	6,012	73,249
31	Carlisle Tire	37716DCTRNMODIN	27.17	2,956	5,010	3,381	10,289	14,314	21,314	57,265
32	General Shale Brick Inc Plant #7	37660GNRLS1740R	36.08					29,222	26,085	55,307
33	Campos Foods	37714PWLLVELKIN	32.88	8,199	10,530	8,050	9,040	6,680	5,203	47,702
34	Benco Sales Inc	38557BNCSL123ST	27.71	5,718	6,020	6,825	7,754	9,030	9,760	45,107
35	Advanced Measurement Technology	37830GGRTC100MI	18.32	3,852	5,854	8,490	8,987	8,045	7,809	43,037
36	Horsehead Corp	37854HRSHDENDOF	9.33	3,285	11,639	5,228	5,179	4,778	5,431	35,540
37	Panasonic Electronic Devices Corp of America	37914MTSSH5105S	39.13			14,594			11,215	25,809
38	PI-INC	37303PNC 2008D	31.69	63	2,570	9,300	6,162	5,723		23,818
39	Suburban Manufacturing Co	37321SBRBN676BR	37.84			11,393			12,353	23,746
40	Denso Manufacturing Athens Tennessee Inc	37303DNSMF24002	29.81	6,048	6,306	3,478	3,265	1,155	1	20,253
41	Goodman Manufacturing Co LP—Dayton	37321GDMNM99MAN	36	2,152	2,600	2,606	2,755	2,542	2,548	15,203

42	Cobalt Yachts LLC	37885MTSSH56EXC	25.17			8,180	4,280			12,460
43	Schott Gemtron	37874GMTRNNEWHW	21.4		21	517	537	11,077	135	12,287
44	Ashland Distribution Knoxville	38106SHLND5263N	39.58		1,087		1,289	3,192	3,288	8,856
45	US Bronze Powders Inc	37801NTDSTROUTE	33.01	542	605	1,144	1,655	1,626	1,300	6,872
46	Custom Quarry	37737CSTMQ566SU	26.73						6,200	6,200
47	Aisin Automotive Casting Tennessee	37716SNTMT221FR	30.94	2,110	2,078	1,107	611	1		5,906
48	Heil Trailer International	37303HLC 1125C	28.52	434	274	187	1,271	1,258	598	4,022
49	US TVA Watts Bar Nuclear Plant	37381STVWT7MILE	26.24	2	7	2,101	753	770	18	3,651
50	Sanford Brands	37801RBBRM1427W	29.41				1,856	1,720		3,576
51	Motiva Enterprises Knoxville Terminal	3792WMTVNT51MID	30.43		3,530					3,530
52	Cymer LLC	37322CYMRL124CY	33.4						3,500	3,500
53	BP Products North America Knoxville Terminal	3792WBPPRD181TH	30.8		2,417					2,417
54	Smokey Mountain Tops Inc	37931SMKYM7216B	26.94				750	750	750	2,250
55	Crossville Inc	38555CRSSV346SW	27.66	628	498	319	220	232	140	2,037
56	Durable Products Inc	38555DRBLP137DA	26.12	250	250	250	250	250	250	1,500
57	Diversified Scientific Services Inc (DSSI)	37763DVRSF657GA	5.44	1	1	4	2	1,398	40	1,446
58	Rocore-Knoxville	37914MDNMN5050S	39.19		250	250	250	250	42	1,042
59	Eagle Bend Manufacturing	37716GLBND1000J	26.74	250	250	250	250			1,000
60	Crossville Inc	38555CRSSV3PRCE	30.18	34	201	180	183	88	88	774
61	Thomas & Betts	37303MRCNLROUTE	31.47	120	120	120	120	120	120	720
62	Tennessee Farmers Cooperative	37853TNNSS4758C	32.88						651	651
63	Polyone Corp	37885VCRNCNILES	26.63				5	511		516
64	Collins & Aikman Corp	37303DVDSN2110C	29.61					250	250	500
65	US DOE East Tennessee Technology Park	37831SDKRDRTE58	7.78	34	104	133	45	0	0	316
66	Energysolutions Bear Creek Facility	37830DRTKB1560B	8.68	22	7	5	4	4	256	299
67	Mayfield Dairy Farms LLC	37303MYFLD813EA	30.16			250				250

(*Continued*)

Table 5.8 TRI Reported Releases to Air for All Industry Within a 40-Mile Radius of the TVA Site—*cont'd*

Ranking	Facility	TRI ID	Distance from Kingston, miles	Total Air Emissions in Pounds						
				2009	2008	2007	2006	2005	2004	Total (2004–2009)
68	Heraeus Metal Processing LLC	37887HRSMT1975K	13.81	3	30	2	5	179		218
69	Cemex Construction Materials Atlantic LLC	37914DXCMN6212C	39.95	10	16	28	30	62	58	205
70	NGK Metals Corp	37874NGKMT917ST	22.14	24	23	22	22	22	22	135
71	Greenkote IPC	37921FNCLR1737L	29.63			10	20	20	20	70
72	A & S Building Systems LP	37714SBLDNHWY11	30.29			13	14	13	15	55
73	Yale Security Inc—Yale Commercial Facility	37771YLSCRYALEA	16.92	9.00	9.00	9.00	9.00	7	7	50
74	Mueller Gas Products Inc	37757LNCLNRT3BO	33.41					25	20	45
75	Langdale Forest Products Co	37874LNGDL555SO	21.15						40	40
76	Modine Manufacturing Co	37716MDNMNJDYAR	28.04			1.00	3.00	4	4	12
77	Greenway Products Inc	37918GRNWY4320G	36.96	5.00	5.00					10
78	Polyone Clinton	37716DHCMP1260C	27.38	1.40	5.00					6.40
79	Knoxville Milling Co	37917WHTLL218EA	34.81					2.43	1.98	4.41
80	Materials & Energy Corp	37830MTRLSETENN	8.54						4.00	4.00
81	Storm Copper Components	37322STRMC240IN	30.97	0.09	0.12	0.12	0.12	0.13	0.13	0.70
82	US DOE Oak Ridge National Laboratory	37831SDKRDBETHE	8.34						0.30	0.30
83	U.S. Natl. Park Service Big South Fork NRRA	37841SNTLP4564L	39.38						0.10	0.10
84	Adams Products—Rockwood	37854BGRCK600CA	5.77	0.00	0.00	0.01	0.01	0.01	0.01	0.05
	Total			4,924,205	12,356,040	14,876,301	14,227,464	15,600,803	15,953,707	77,938,520

Source: Data retrieved from TOXNET URL—http://toxnet.nlm.nih.gov/. See TRI Query (where you can input facility information to retrieve TRI reports): http://www.epa.gov/enviro/html/tris/tris_query.html

Table 5.9 Percent Contribution to Air Pollution from Each Industrial Facility and Change in Reported Releases to Air Between 2008 and 2009

Ranking	Facility	TRI ID	Distance from Kingston (miles)	Percent of Total Regional TRI Reported Air Emissions Between 2004–2009		Percent Change in Reported Releases to Air Between 2008–2009
				Total Regional	Cumulative	
1	US TVA Kingston Fossil Plant	37763STVKNSWANP	—	32	32	-74
2	US TVA Bull Run Fossil Plant	37716STVBL1265E	22.77	24	56	-91
3	Viskase Corp	37774VSKSCEASTL	15.70	15	72	1.3
4	ALCOA	37701LMNMCPOBOX	32.09	10	81	-60
5	Sea Ray Boats Tellico Facility	37885SRYBT100SE	24.70	3	84	-65
6	Sea Ray Boats Knoxville Facility	37914SRYBT2601S	38.86	2	84	-70
7	Tate & Lyle Loudon	37774STLYMROUTE	16.41	2	88	-14.7
8	Skier's Choice Inc	37801SCHLG1717H	30.17	2	90	-62.7
9	Tennessee Watercraft Inc	37885CBBTC2000C	25.28	1	91	-62
10	Mastercraft Boat Co	37885MSTRC100CH	24.73	1	93	-77.5
11	NNSA Y-12 National Security Complex	37831SDKRDBEARC	16.78	1	94	72
12	American Safety Razor Co LLC	37914MRCNS2820M	39.84	1	94	-56.7
13	Malibu Boats LLC	37774MLBBTHIGHL	14.70	1		-42
14	Acupowder TN LLC	37742GRNBCBOX63	27.77	1		-84
15	Rohm & Haas Chemicals LLC Knoxville Site	37921RHMND730DA	33.53	0		3.8
16	Johns Manville	37331MNVLLHWY41	36.19	0		-11.6

(*Continued*)

Table 5.9 Percent Contribution to Air Pollution from Each Industrial Facility and Change in Reported Releases to Air Between 2008 and 2009—*cont'd*

Ranking	Facility	TRI ID	Distance from Kingston (miles)	Percent of Total Regional TRI Reported Air Emissions Between 2004–2009: Total Regional	Percent of Total Regional TRI Reported Air Emissions Between 2004–2009: Cumulative	Percent Change in Reported Releases to Air Between 2008–2009
17	Hubbell Lenoir City Inc	37771QZTCR300IN	16.82	0		-13
18	Thyssenkrupp Waupaca Plant 6	37331THYSS134WA	36.85	0		-37
19	Huber Engineered Woods LLC Whites Creek Facility	37381JMHBR1261D	17.37	0		-100
20	Kimberly-Clark Corp Loudon Mill	37774KMBRL5600K	13.97	0		-60
21	Ficosa NA	38555DLBRP87WYA	27.74	0		-100
22	Denso Manufacturing Tennessee Inc	37801NPPND1720R	31.13	0		-12
23	Omega Cabinetry	37716HMCRSHWY25	27.14	0		-100
24	Vestal Manufacturing Ent. Inc.	37874VSTLM177IN	20.46	0		-30
25	Toho Tenax America Inc	37854FRTFLROANE	6.78	0		-79
26	Gerdau Ameristeel Knoxville	37921FLRDS1919T	32.80	0		-2.7
27	Becromal of America	37716BCRML350JD	28.51	0		3.5
28	United States Utility Enclosures	37874NTDST223HW	23.74	0		1.2
29	La-Z-Boy Tennessee	37321LZBYTWALNU	36.66	0		13
30	Aleris Recycling Inc	37774MTLRSBLAIR	16.49	0		427
31	Carlisle Tire	37716DCTRNMODIN	27.17	0		-41

32	General Shale Brick Inc Plant #7	37660GNRLS1740R	36.08	0	
33	Campos Foods	37714PWLLVELKIN	32.88	0	-22
34	Benco Sales Inc	38557BNCSL123ST	27.71	0	-5
35	Advanced Measurement Technology	37830GGRTC100MI	18.32	0	-34
36	Horsehead Corp	37854HRSHDENDOF	9.33	0	-71
37	Panasonic Electronic Devices Corp	37914MTSSH5105S	39.13	0	
38	PI-INC	37303PNC 2008D	31.69	0	-97
39	Suburban Manufacturing Co	37321SBRBN676BR	37.84	0	
40	Denso Manufacturing Athens TN	37303DNSMF24002	29.81	0	-4
41	Goodman Manufacturing Co LP—Dayton	37321GDMNM99MAN	36.00	0	-17
42	Cobalt Yachts LLC	37885MTSSH56EXC	25.17	0	
43	Schott Gemtron	37874GMTRNNEWHW	21.40	0	-100
44	Ashland Distribution Knoxville	38106SHLND5263N	39.58	0	-100
45	US Bronze Powders Inc	37801NTDSTROUTE	33.01	0	-10
46	Custom Quarry	37737CSTMQ566SU	26.73	0	
47	Aisin Automotive Casting Tennessee	37716SNTMT221FR	30.94	0	1.5
48	Heil Trailer International	37303HLC 1125C	28.52	0	58
49	US TVA Watts Bar Nuclear Plant	37381STVWT7MILE	26.24	0	-69
50	Sanford Brands	37801RBBRM1427W	29.41	0	
51	Motiva Ent. Knoxville Terminal	3792WMTVNT51MID	30.43	0	-100
52	Cymer LLC	37322CYMRL124CY	33.40	0	

(Continued)

Table 5.9 Percent Contribution to Air Pollution from Each Industrial Facility and Change in Reported Releases to Air Between 2008 and 2009—*cont'd*

Ranking	Facility	TRI ID	Distance from Kingston (miles)	Percent of Total Regional TRI Reported Air Emissions Between 2004–2009		Percent Change in Reported Releases to Air Between 2008–2009
				Total Regional	Cumulative	
53	BP Products North America Knoxville Terminal	3792WBPPRD181TH	30.80	0		-100
54	Smokey Mountain Tops Inc	37931SMKYM7216B	26.94	0		
55	Crossville Inc	38555CRSSV346SW	27.66	0		26
56	Durable Products Inc	38555DRBLP137DA	26.12	0		0
57	Diversified Scientific Services Inc	37763DVRSF657GA	5.44	0		-39
58	Rocore-Knoxville	37914MDNMN5050S	39.19	0		-100
59	Eagle Bend Manufacturing	37716GLBND1000J	26.74	0		0
60	Crossville Inc	38555CRSSV3PRCE	30.18	0		-83
61	Thomas & Betts	37303MRCNLROUTE	31.47	0		0
62	Tennessee Farmers Cooperative	37853TNNSS4758C	32.88	0		
63	Polyone Corp	37885VCRNCNILES	26.63	0		
64	Collins & Aikman Corp	37303DVDSN2110C	29.61	0		
65	US DOE East Tennessee Technology Park	37831SDKRDRTE58	7.78	0		-66
66	Energysolutions Bear Creek Facility	37830DRTKB1560B	8.68	0		234
67	Mayfield Dairy Farms LLC	37303MYFLD813EA	30.16	0		
68	Heraeus Metal Processing LLC	37887HRSMT1975K	13.81	0		-91
69	Cemex Construction Materials	37914DXCMN6212C	39.95	0		-39
70	NGK Metals Corp	37874NGKMT917ST	22.14	0		4

71	Greenkote IPC	37921FNCLR1737L	29.63	0	
72	A & S Building Systems LP	37714SBLDNHWY11	30.29	0	
73	Yale Security Inc	37771YLSCRYALEA	16.92	0	0
74	Mueller Gas Products Inc	37757LNCLNRT3BO	33.41	0	
75	Langdale Forest Products Co	37874LNGDL555SO	21.15	0	
76	Modine Manufacturing Co	37716MDNMNJDYAR	28.04	0	
77	Greenway Products Inc	37918GRNWY4320G	36.96	0	0
78	Polyone Clinton	37716DHCMP1260C	27.38	0	
79	Knoxville Milling Co	37917WHTLL218EA	34.81	0	
80	Materials & Energy Corp	37830MTRLSETENN	8.54	0	
81	Storm Copper Components	37322STRMC240IN	30.97	0	
82	US DOE Oak Ridge National Laboratory	37831SDKRDBETHE	8.34	0	
83	U.S. Natl. Park Service Big South Fork NRRA	37841SNTLP4564L	39.38	0	
84	Adams Products—Rockwood	37854BGRCK600CA	5.77	0	
			Total	**100**	**-60**

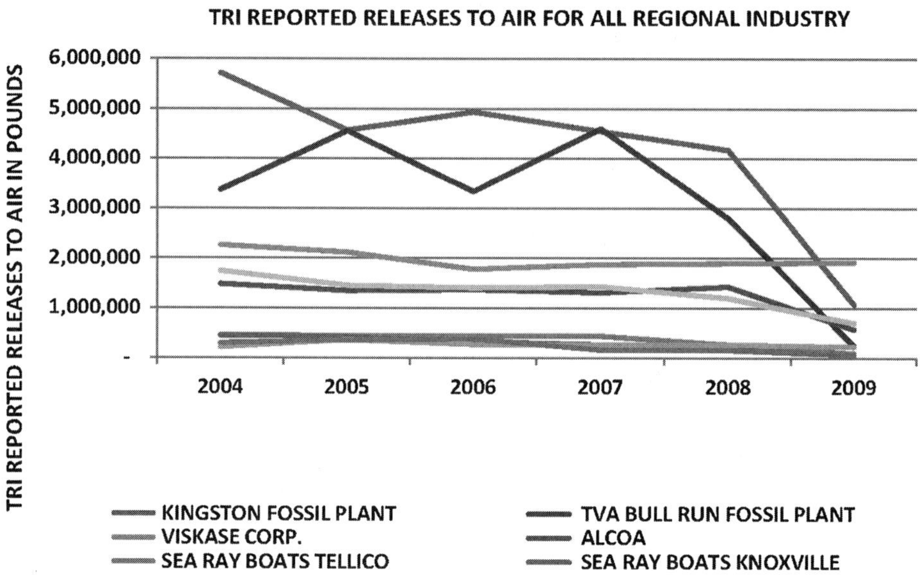

Figure 5.16 TRI reported releases to air by facility.

reductions in releases to air between 2008 and 2009. In fact, KIF reported more than 74% fewer emissions in 2009 than in 2008, and the Bull Run Fossil Plant reported 91.5% fewer releases to air in 2009. Both of these fossil-burning plants generated less electricity, burned less coal, and released fewer emissions to air. KIF burned 80% less coal in 2009 versus 2008 (see Tables 5.10A and 5.10B). But in addition the entire regional industrial community reported 60% fewer releases to air in 2009 compared to 2008. Air monitoring conducted by TVA and its contractors throughout 2009 do not even come close to agreeing with industry-reported reductions in emissions to the region. It is illogical for 60% fewer emissions to air to cause only a marginal improvement to air quality.

The TRI report releases for certain chemicals that the EPA has identified as hazardous air pollutants. Under the TRI program, facilities are not required to report particulate matter. PM is, however, recognized by the EPA as a health hazard. According to the EPA,

> *Particulate matter, also known as particle pollution or PM, is a complex mixture of extremely small particles and liquid droplets. Particle pollution is made up of a number of components, including acids (such as nitrates and sulfates), organic chemicals, metals, and soil or dust particles. The size of particles is directly linked to their potential for causing health problems. EPA is concerned about particles that are 10 micrometers in*

Table 5.10A Power Production Statistics for KIF[*]

Year	MWh	Source
2003	10,120,626	http://www.tva.gov/environment/air/kingston_tri_archive.htm
2004	9,690,784	http://www.tva.gov/environment/air/kingston_tri_archive.htm
2005	9,479,726	http://www.dirtykilowatts.org/Dirty_Kilowatts.pdf
2006	10,377,572	http://www.dirtykilowatts.org/Dirty_Kilowatts2007.pdf
2007	ND	
2008	11,066,325	http://webcache.googleusercontent.com/search?q=cache: V9SnYE4EW3IJ:www.tva.gov/environment/air/kingston. htm+2009+tri+kingston+fossil+tva+Gross+MWh+generated&; cd=1&hl=en&ct=clnk&gl=us
2009	2,141,325	http://www.tva.gov/environment/air/kingston.htm

[*]Shows KIF produced only 19% as much electricity in 2009 compared to 2008. KIF burned 80% less coal.

Table 5.10B Public Records of Electricity Generation for TVA Bull Run[*]

Year	MWH	Source
2003	6,443,620	http://www.tva.gov/environment/air/bullrun_tri_archive.htm
2004	5,089,880	http://www.tva.gov/environment/air/bullrun_tri_archive.htm
2005	6,587,568	http://www.dirtykilowatts.org/Dirty_Kilowatts.pdf
2006	4,696,141	http://www.dirtykilowatts.org/Dirty_Kilowatts2007.pdf
2008	4,333,294	http://webcache.googleusercontent.com/search?q=cache: 2KT5Ywy7tMAJ:www.tva.gov/environment/air/bullrun. htm+tva+bull+run+Gross+MWh+generated&; cd=1&hl=en&ct=clnk&gl=us
2009	3,524,611	https://www.tva.gov/environment/air/bullrun.htm#tri

[*]Shows TVA Bull Run Fossil Plant produced 81% as much electricity in 2009 compared to 2008. Additionally, a new scrubber was installed in late 2008.

diameter or smaller because those are the particles that generally pass through the throat and nose and enter the lungs. Once inhaled, these particles can affect the heart and lungs and cause serious health effects. EPA groups particle pollution into two categories: (1) "Inhalable coarse particles," such as those found near roadways and dusty industries, are larger than 2.5 micrometers and smaller than 10 micrometers in diameter; (2) "Fine particles," such as those found in smoke and haze, are 2.5 micrometers in diameter and smaller. These particles can be directly emitted from sources such as forest fires, or they can form when gases emitted from power plants, industries and automobiles react in the air.

<http://www.epa.gov/pm/>

For this very reason, the EPA instructed TVA to monitor PM in the region when the ash spill occurred. TVA, in fact, monitored PM at a number of locations and reported that between 2008 and 2009 the annual average $PM_{2.5}$ concentration for the regional area and Harriman dropped by approximately 18.5% and 19.5%, respectively. There are no reported values for annual average $PM_{2.5}$ for KIF prior to 2009. This shows that the air quality in the region seemingly improved between 2008 and 2009.

The data that TVA prepared on regional PM concentration and presented to the public are less than candid because they provide the wrong impression that air quality in the communities in proximity to the spill site improved during the time period following the spill incident. My review of the air quality data and industrial activities of the region show that air quality actually worsened with respect to PM. As stated previously, facilities are not required to report PM releases to the TRI; however, they are required to report such releases to the National Emissions Inventory. The EPA's National Emissions Inventory reports point source (stack) emissions every 3 years. The first inventory was prepared in 1999. The 2008 NEI report has not been published by the EPA, so I have relied on the most recent NEI data reported in 2005 (see Figure 5.17). These data were assembled from EPA's website for those facilities that account for 90% of the reported releases to air according to the TRI reporting data summarized

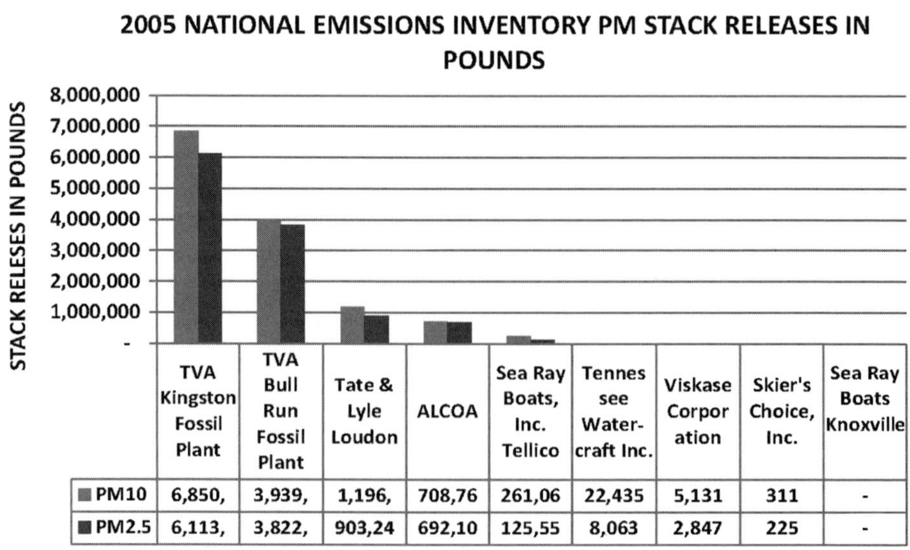

	TVA Kingston Fossil Plant	TVA Bull Run Fossil Plant	Tate & Lyle Loudon	ALCOA	Sea Ray Boats, Inc. Tellico	Tennes see Water-craft Inc.	Viskase Corpor ation	Skier's Choice, Inc.	Sea Ray Boats Knoxville
PM10	6,850,	3,939,	1,196,	708,76	261,06	22,435	5,131	311	-
PM2.5	6,113,	3,822,	903,24	692,10	125,55	8,063	2,847	225	-

Figure 5.17 NEI inventory of stack releases.
Source: NEI URL—http://www.epa.gov/ttn/chief/eiinformation.html; NEI 2005 URL—http://www.epa.gov/ttn/chief/net/2005inventory.html

Table 5.11 Percent Contribution of PM Stack Releases by Facility

Facility	2005 NEI Reported Releases in Pounds		Stack Heights	% of Total PM_{10}
	PM_{10}	PM_{25}		
TVA Kingston Fossil Plant	6,850,064	6,113,214	1000 ft	52.76
TVA Bull Run Fossil Plant	3,939,124	3,822,010	800 ft	30.34
Tate & Lyle Loudon	1,196,633	903,243	Short	9.22
ALCOA	708,769	692,109	Short	5.46
Sea Ray Boats, Inc. Tellico	261,060	125,558	Short	2.01
Tennessee Watercraft Inc.	22,435	8,063	Short	0.17
Viskase Corporation	5,131	2,847	Short	0.04
Skier's Choice, Inc.	311	225	Short	0.00
Sea Ray Boats Knoxville	—	—	Short	—
Total	**12,983,527**			**100.00**

in Table 5.8. The EPA's data were assembled and are presented in Figures 5.18A through C and in Table 5.11.

The EPA's data show that KIF is the largest emitter of PM in the region. When the plant operates under normal conditions, it releases more than 6.8 million pounds of PM_{10}, accounting for 53% of the particulate matter released

Figure 5.18A Location and annual mean PM in micrograms per cubic meter for site locations 47-145-1001 and 47-145-3009.

into the region. The TVA Bull Run Plant contributes another 3.9 million pounds, or 30%. Combined, the two TVA plants account for more than 80% of the PM releases into the region. These releases are from 1,000-foot-tall stacks from KIF and 800-foot-tall stacks from the Bull Run Fossil Plant. In 2009 both of these facilities reported huge reductions in stack emissions. KIF reported more than 74% fewer emissions in its 2009 TRI report than in 2008, and the Bull Run Fossil Plant reported 91.5% fewer releases to air in 2009. Both of these fossil-burning plants generated less electricity, burned less coal, and released fewer emissions to air. In fact, KIF burned 80% less coal in 2009 versus 2008.

Such significant reductions in PM releases from stacks are inconsistent with the PM air monitoring results reported by TVA for 2009. It is illogical that 74%–91% reductions in stack emissions for two facilities (KIF and Bull Run), which account for more than 80% of the PM releases into the region, would have only a marginal improvement to air quality.

Regional air pollution emissions in 2009 were reduced overall by 60% in 2009 compared to 2008. Such a large reduction in pollution loadings is in disagreement with the marginal reductions in PM monthly averages that TVA has reported from its air monitoring stations. Such large reductions in pollution loadings would cause nearly comparable reductions in PM, and yet this is not what the monitoring stations show. This observation means that there was another source of PM influencing the monthly average concentrations. There is no reason to assume that mobile sources of emissions dramatically increased between the two years; and there are no volcanoes in Tennessee. The single event that can explain the fact that regional air quality remained stressed throughout 2009 is the release of 1.1 billion gallons of coal fly ash slurry onto the surrounding land area.

Further evidence that the regional air quality has been impacted by the spill was obtained from a review of air monitoring data obtained from EPA. The regional EPA office provided a download of all the PM monitoring data that it collected in the state of Tennessee in 2009. Although there is a large number of PM monitoring stations in the state, there were only seven PM_{10} monitoring stations within the impacted region.

Figure 5.18A shows the location and the annual mean PM_{10} concentration for two locations. These two stations are essentially on the fence line of the TVA property or close to it. The EPA reported values are between 17 to nearly 18 micrograms per cubic meter. Figure 5.18B shows the location and annual mean PM_{10} concentrations for two monitoring stations located about 10 miles southwest of the TVA site. These two monitoring stations show PM_{10} concentrations that are on the average more than 14% lower than the concentrations measured close to the TVA site. Figure 5.18C shows the

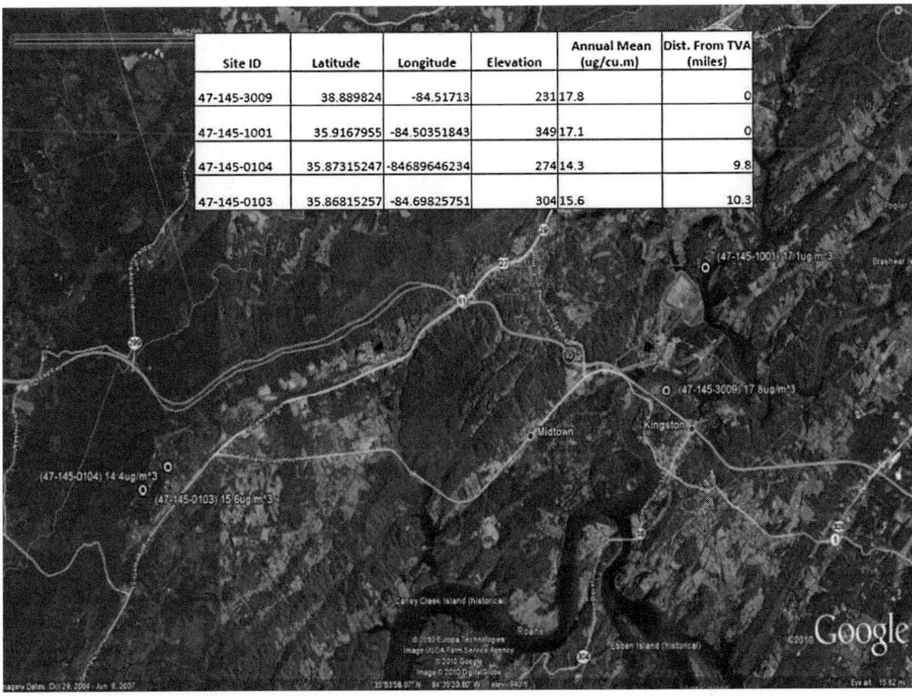

Site ID	Latitude	Longitude	Elevation	Annual Mean (ug/cu.m)	Dist. From TVA (miles)
47-145-3009	38.889824	-84.51713	231	17.8	0
47-145-1001	35.9167955	-84.50351843	349	17.1	0
47-145-0104	35.87315247	-84689646234	274	14.3	9.8
47-145-0103	35.86815257	-84.69825751	304	15.6	10.3

Figure 5.18B Location and annual mean PM in for site locations 47-145-0104 AND 47-145-0103

location of the three remaining stations, which are located about 34 miles away in Knoxville. Two of these stations reported annual mean concentrations that were between 12% and 17% lower, while one was about 5% higher than the data reported by two monitoring stations closest to the TVA site. On the average, the annual mean PM_{10} for the Knoxville area was about 8% lower than the stations closest to the TVA site. Further, the stations in Knoxville are within the most industrialized zone of the region, yet they still showed lower PM_{10} concentrations than recorded near the TVA fence line. The EPA's data show that the area around the TVA facility has elevated PM levels.

Further evidence that the region has been impacted by ash from the spill was gathered from air sampling conducted using a DustTrak II unit and sampling pumps that collected airborne particulate matter (see Figure 5.19). Airborne dust samples collected in the filters of the DustTrak II unit were also analyzed to determine their chemical composition. The laboratory results showed detectable levels of arsenic, barium, beryllium, cadmium, chromium, copper, lead, selenium, vanadium, zinc, and radionuclides. All of these

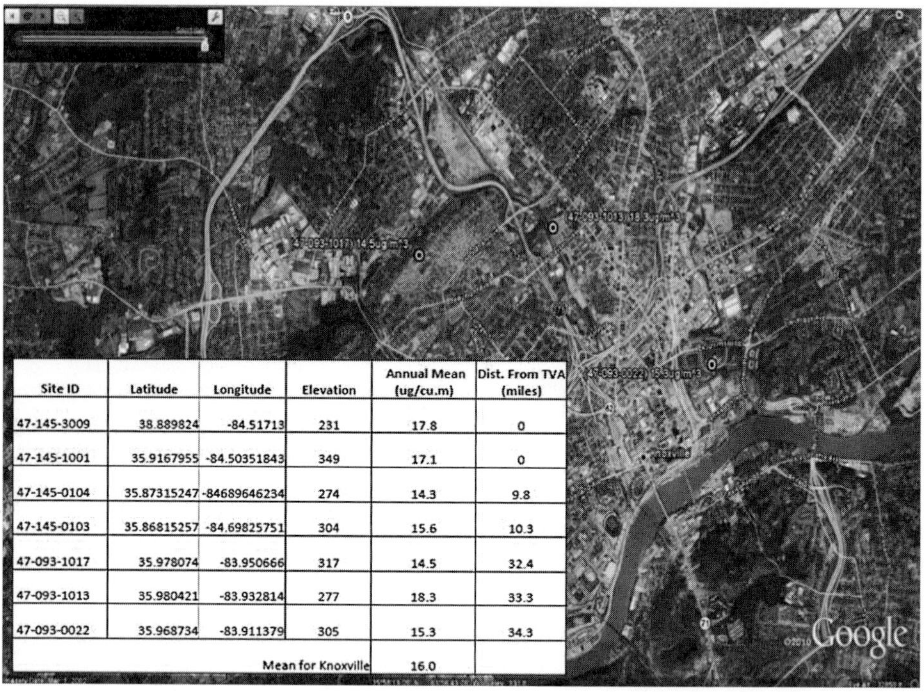

Site ID	Latitude	Longitude	Elevation	Annual Mean (ug/cu.m)	Dist. From TVA (miles)
47-145-3009	38.889824	-84.51713	231	17.8	0
47-145-1001	35.9167955	-84.50351843	349	17.1	0
47-145-0104	35.87315247	-84689646234	274	14.3	9.8
47-145-0103	35.86815257	-84.69825751	304	15.6	10.3
47-093-1017	35.978074	-83.950666	317	14.5	32.4
47-093-1013	35.980421	-83.932814	277	18.3	33.3
47-093-0022	35.968734	-83.911379	305	15.3	34.3
			Mean for Knoxville	16.0	

Figure 5.18C Location and annual mean PM concentrations in Knoxville.

Figure 5.19 Locations of air sampling using DustTrak and sampling pumps.

chemical toxins are reported in authoritative studies to be found in coal ash. TVA's Material Safety Data Sheets and literature on its website report and acknowledge the presence of such toxins in its fly and bottom ashes. These results are an independent confirmation that fly and bottom ashes were airborne and off-site.

The preceding information and data collected support that the finding that ash from the spill traveled off-site into air surrounding the TVA spill. Using the NEI reported stack releases of PM, reductions of pollution loadings to the region reported by facilities to the NEI, and TVA's reported reduction in monthly average PM concentrations from its air monitoring stations in 2009, an estimate of the airborne ash was made. This estimate shows that approximately 6.1 million pounds of PM_{10} (containing approximately 5.4 million pounds of $PM_{2.5}$) were released into the region's air as a result of the ash spill.

The calculations for the estimate are provided in Table 5.12. These releases were fugitive emissions. The particulate matter released was fly and bottom ash. While the monthly average PM for the region was lower in 2009 compared to 2008 by a nominal 20%, the region was exposed to vastly greater fugitive emissions than any year prior to the history of the ash spill. These fugitive releases were from ground level and not 1,000-foot-tall stacks. The release of these emissions exposed the region and its community members to a wide range of toxic chemicals.

The following explains the calculated values reported in Table 5.12:

1. Columns (1) and (2) tabulate the 2005 NEI stack releases for each plant in pounds.
2. TVA reports the annual average PM_{10} conc. ($\mu g/m^3$). For 2005, this value was $25\mu g/m^3$. Since we are comparing the year 2008 to 2009, first we need to know the stack emissions in 2008. TVA reported that the air quality in 2008 was $22.5\mu g/m^3$. Therefore, the percent change in annual average PM_{10} conc. was $100 \times (22.5 - 25)/25 = -10\%$ for PM_{10}. Similarly, in 2005, $PM_{2.5} = 15.2\ \mu g/m^3$ and in 2008, $PM_{2.5} = 13\ \mu g/m^3$. Therefore, the percent change between 2005 and 2008 = $100\ (13 - 15.2)/15.2 = -14.5\%$.
3. Columns (3) and (4) are the calculated stack releases for PM_{10} and $PM_{2.5}$ for 2008, respectively. These are calculated by assuming that the reductions in regional PM reported by Wylie between 2005 and 2008 were due to reductions in facility stack releases. Since KIF and Bull Run represent 80% of the PM releases into the region, this is a reasonable assumption. As a sample calculation for KIF, PM_{10} stack releases in 2008 = $(1 - 0.10) \times 6{,}850{,}064 = 6{,}165{,}058$ pounds in PM_{10}. Similarly, for $PM_{2.5}$ for KIF = $(1 - 0.145) \times 6{,}113{,}214 = 5{,}228{,}407$ pounds (see column 4). These calculations were repeated for each facility with values tabulated in columns (3) and (4).
4. Values tabulated in columns (3) and (4) were totaled. See column (3): the total PM_{10} for the nine facilities is 11,681,174 pounds PM_{10}. See column (4): the total

Table 5.12 Calculations Summary

Row	Facility	2005 NEI Stack Releases (Pounds)		2008 Stack Emissions			2009 Stack Releases (Pounds)	
		Column (1) PM_{10}	**Column (2) PM_{25}**	**Column (3) PM_{10} (Pounds)**	**Column (4) $PM_{2.5}$ (Pounds)**	**Column (5) % Change in Releases to Air 2008–2009**	**Column (6) PM_{10}**	**Column (7) PM_{25}**
1	TVA Kingston Fossil Plant	6,850,064	6,113,214	6,165,058	5,228,407	(74)	1,602,915	1,359,386
2	TVA Bull Run Fossil Plant	3,939,124	3,822,010	3,545,212	3,268,824	(91)	319,069	294,194
3	Tate & Lyle Loudon	1,196,633	903,243	1,076,970	772,510	(15)	918,655	658,951
4	ALCOA	708,769	692,109	637,892	591,935	(60)	255,157	236,774
5	Sea Ray Boats, Inc. Tellico	261,060	125,558	234,954	107,385	(65)	82,234	37,585
6	Tennessee Watercraft Inc.	22,435	8,063	20,192	6,896	(62)	7,673	2,620
7	Viskase Corporation	5,131	2,847	4,618	2,435	1	4,678	2,467
8	Skier's Choice, Inc.	311	225	280	192	(63)	104	72
9	Sea Ray Boats Knoxville	—	—	—	—	(70)	—	—
10	Total	12,983,527	11,667,269	11,685,174	9,978,585		3,190,485	2,592,049
11	2009 Expected Stack Releases (Pounds)	9,348,140	7,982,868					
12	Fugitive Emissions from ash spill (Pounds)	6,157,654	5,390,819					

$PM_{2.5}$ for all nine facilities is 9,978,585 pounds. These values are the 2008 PM stack releases for all nine facilities.

5. According to TVA, PM air quality (concentration) was reduced by about 20% between 2008 and 2009. Therefore, the 2009 stack releases should be 80% of the 2008 releases. Hence, for PM_{10}, the total stack releases for 2009 = (1 – 0.20) × 11,685,174 = 9,348,139 pounds. Similarly, the total $PM_{2.5}$ releases are = (1 – 0.2) × 9,978,585 = 7,982,868 pounds. These values are shown in row (11).

6. Columns (6) and (7) report the stack PM_{10} and $PM_{2.5}$ releases for 2009, respectively. These are calculated from the TRI reported reductions in releases to air (see percentages in column [5]). As a sample calculation for KIF, the 2009 PM_{10} stack release was 74% lower than 2008 because that is what was reported by KIF to the TRI. Hence, PM_{10} for 2009 = (1 – 0.74) × 6,165,058 = 1,602,915 pounds. Similarly, for $PM_{2.5}$ = (1 – 0.74) × 5,228,407 = 1,359,386 pounds. These computations are repeated for all nine facilities, and the totals are reported at the bottom of columns (6) and (7).

7. The mass of fugitive emissions released to air is the difference between the expected stack releases (row 11) and the 2009 stack releases. Hence, for PM_{10} = 9,348,140 – 3,190,485 = 6,157,654 pounds released to air. Similarly, for $PM_{2.5}$ = 7,982,868 – 2,592,049 = 5,390,819 pounds released to air. Note that $PM_{2.5}$ is a subset of PM_{10} (they are not additive).

Fly ash and bottom ash are byproducts of coal combustion. Fly ash is captured by the pollution controls used at KIF and slurried into ponds along with pulverized bottom ash. Coal bottom ash is a coarse, granular byproduct of coal combustion that is collected from the bottom of coal furnaces in power plants. When pulverized coal is burned in a dry bottom boiler, approximately 80% or more of the unburned material is captured from the chimney as fly ash. The remaining 20% is bottom ash, which is collected in a water-filled hopper at the bottom of the furnace. These ashes contain mercury, arsenic, beryllium, cadmium, chromium, nickel, selenium, and many other toxic metals that were confirmed by sampling using the DustTrak II. When coal is burned, toxins in the coal are released into the furnace and smokestack. With modern air pollution controls, airborne toxins are captured through filtration and scrubber systems before they can become airborne, and are contained in a fine ash called coal ash, fly ash, or coal combustion waste. As a result, heavy metals such as mercury are concentrated in what the EPA considers "recycled air pollution control residue." These same byproducts of coal combustion contain radioactive materials. Zielinski and Budahn[17] as well as many other researchers have applied radiation-based techniques for determining the distribution and relative abundance of radionuclides found in fly ash and bottom ash samples

[17] Robert A. Zielinski and James R. Budahn, "Radionuclides in Fly Ash and Bottom Ash: Improved Characterization Based on Radiography and Low Energy Gamma-Ray Spectrometry," *Fuel,* March 1998, 77(4): 259–267.

from coal-fired power plants. Such researchers have reported observations of the variety of uranium hosts and of uranium distribution in individual particles of fly and bottom ashes. They have reported on the relative abundance of ^{238}U, ^{226}Ra, ^{228}Ra, and ^{210}Pb in ash samples. Results indicate that radium isotopes are not significantly (within 10%–15%) fractionated from parent ^{238}U and ^{232}Th during coal combustion. In contrast, ^{210}Pb appears to be preferentially enriched in some samples of fly ash and depleted in bottom ash relative to ^{238}U and ^{226}Ra. These toxic materials were released as a part of the 6.1 million pounds of fugitive particulate matter into the region.

The analysis supports that more than 6 million pounds of particulate matter containing toxic metals and radionuclides were released at ground level in the form of fugitive dust emissions throughout 2009. Nearly half of the industrial community reported substantial reductions in releases to air in 2009, and eight out of nine of the top polluters that account for 90% of the reported air emissions in the region reported significant reductions of releases to air in their 2009 TRI reporting. Only one major polluter reported a 1% increase in air releases, which is a tiny fraction of the PM releases to the region. KIF air emissions have historically been the greatest, and TVA's own monitoring data show that monthly average PM concentrations are directly influenced by the facility's stack emissions.

5.12 Industry-Wide Problem

Table 5.13 reports the power plants in each state with coal ash ponds and the amount in tons stored, according to an Associated Press analysis of Energy Department data from 2005, the latest year statistics were available. In total, there are 156 waste ponds with a combined capacity of 19,699,130 tons. These waste management units have not been fully surveyed and assessed for

Table 5.13 List of Ash Ponds in the United States

State	Facility Name	County	Capacity, Tons	Percent of Reported Capacity
Alabama	Alabama Electric Cooperative Inc.	Washington	28,400	0.14
	Alabama Power Co.	Mobile	282,900	1.44
	Alabama Power Co.	Etowah	34,100	0.17
	Alabama Power Co.	Walker	304,900	1.55
	Alabama Power Co.	Greene	211,900	1.08
	Alabama Power Co.	Jefferson County	61,500	0.31

Table 5.13 List of Ash Ponds in the United States—*cont'd*

State	Facility Name	County	Capacity, Tons	Percent of Reported Capacity
	Tennessee Valley Authority	Colbert	29,200	0.15
	Tennessee Valley Authority	Jackson	407,600	2.07
Arizona	Arizona Electric Power Cooperative Inc.	Cochise	33,000	0.17
	Arizona Public Service Co.	Navajo	258,000	1.31
Arkansas	Domtar Industries Inc.	Little River	40,300	0.20
	Southwestern Electric Power Co	Benton	19,400	0.10
Colorado	Platte River Power Authority	Larimer	5,700	0.03
Florida	Gulf Power Co.	Bay	70,300	0.36
	Tampa Electric Co.	Hillsborough	200	0.00
Georgia	Georgia Power Co.	Bartow	93,300	0.47
	Georgia Power Co.	Putnam	416,300	2.11
	Georgia Power Co.	Heard	536,700	2.72
	Georgia Power Co.	Monroe	470,600	2.39
	Savannah Electric & Power Co.	Chatham	10,000	0.05
	Savannah Electric & Power Co.	Effingham	15,000	0.08
Illinois	Ameren Energy Generating Co.	Crawford	31,000	0.16
	Ameren Energy Generating Co.	Morgan	48,000	0.24
	Ameren Energy Generating Co.	Jasper	109,000	0.55
	Ameren Energy Resources Generating	Peoria	52,000	0.26
	Ameren Energy Resources Generating	Fulton	63,000	0.32
	Dynegy Midwest Generation Inc.	Randolph	116,000	0.59
	Dynegy Midwest Generation Inc.	Mason County	86,000	0.44
	Dynegy Midwest Generation Inc.	Putnam	20,800	0.11
	Dynegy Midwest Generation Inc.	Vermilion	13,700	0.07
	Dynegy Midwest Generation Inc.	Madison	14,200	0.07
	Springfield City of	Sangamon	72,100	0.37
Indiana	Alcoa Power Generating Inc.	Warrick	241,900	1.23
	Hoosier Energy R E C Inc.	Pike	39,800	0.20
	Indiana-Kentucky Electric Corp.	Jefferson	21,700	0.11
	Indianapolis Power & Light Co.	Marion	175,900	0.89
	Indiana Michigan Power Co.	Dearborn	140,600	0.71
	Indiana Michigan Power Co.	Spencer	11,800	0.06
	PSI Energy Inc.	Vermillion	210,900	1.07
	PSI Energy Inc.	Knox	11,500	0.06
	PSI Energy Inc.	Floyd	125,600	0.64
	PSI Energy Inc.	Vigo	192,100	0.98
	PSI Energy Inc.	Gibson	897,800	4.56
	Southern Indiana Gas & Electric Co.	Warrick	35,600	0.18
	Southern Indiana Gas & Electric Co.	Posey	165,750	0.84
Iowa	Interstate Power & Light Co.	Allamakee	24,000	0.12

(*Continued*)

Table 5.13 List of Ash Ponds in the United States—*cont'd*

State	Facility Name	County	Capacity, Tons	Percent of Reported Capacity
	MidAmerican Energy Co.	Pottawattamie	104,500	0.53
	MidAmerican Energy Co	Woodbury	50,200	0.25
	MidAmerican Energy Co.	Louisa	23,000	0.12
Kansas	Kansas City	Wyandotte	10,200	0.05
	Westar Energy	Pottawatomie	184,100	0.93
Kentucky	Cincinnati Gas & Electric Co.	Boone	172,900	0.88
	East Kentucky Power Cooperative Inc.	Clark	60,000	0.30
	East Kentucky Power Cooperative Inc.	Mason	4,300	0.02
	Kentucky Utilities Co.	Mercer	140,500	0.71
	Kentucky Utilities Co.	Carroll	634,700	3.22
	Kentucky Utilities Co.	Muhlenberg	30,600	0.16
	Kentucky Utilities Co.	Woodford	18,900	0.10
	Louisville Gas & Electric Co.	Jefferson	37,100	0.19
	Louisville Gas & Electric Co.	Jefferson	64,700	0.33
	Louisville Gas & Electric Co.	Trimble	150,900	0.77
	Tennessee Valley Authority	Muhlenberg	125,700	0.64
	Tennessee Valley Authority	McCracken	61,100	0.31
	Western Kentucky Energy Corp.	Henderson	12,300	0.06
	Western Kentucky Energy Corp.	Webster	21,800	0.11
	Kentucky Power Co.	Lawrence	298,300	1.51
Louisiana	Cleco Power LLC	De Soto Parish	51,900	0.26
	Louisiana Generating LLC	Pointe Coupee Parish	139,400	0.71
Maryland	Mirant Mid-Atlantic LLC	Montgomery	3,000	0.02
	Allegheny Energy Supply Co. LLC	Washington	25,100	0.13
Michigan	Consumers Energy Co.	Bay	108,800	0.55
	Consumers Energy Co.	Bay	69,900	0.35
	Consumers Energy Co.	Monroe	3,400	0.02
	Detroit Edison Co.	Monroe	482,000	2.45
	Lansing City of	Eaton	5,100	0.03
Minnesota	Allete Inc.	St. Louis County	20,200	0.10
	Allete Inc.	Itasca	163,400	0.83
	Northern States Power Co.	Dakota	4,800	0.02
	Northern States Power Co.	Ramsey	10	0.00
	Northern States Power Co.	Hennepin	6,700	0.03
	Northern States Power Co.	Sherburne	355,700	1.81
Mississippi	Mississippi Power Co.	Harrison	39,100	0.20
	Weyerhaeuser Co.	Lowndes	60,000	0.30
Missouri	Associated Electric Cooperative Inc.	New Madrid	109,200	0.55

Table 5.13 List of Ash Ponds in the United States—*cont'd*

State	Facility Name	County	Capacity, Tons	Percent of Reported Capacity
	Empire District Electric Co.	Jasper	53,500	0.27
	Independence City of	Jackson	29,750	0.15
	Kansas City Power & Light Co.	Platte	16,400	0.08
	Sikeston City of	Scott	11,300	0.06
	Ameren UE	Franklin	250,000	1.27
	Ameren UE	St. Louis	111,000	0.56
	Ameren UE	St. Charles	102,000	0.52
	Ameren UE	Jefferson	96,000	0.49
Montana	PPL Montana LLC	Rosebud	963,600	4.89
New Mexico	Arizona Public Service Co.	San Juan	461,700	2.34
North Carolina	Progress Energy Carolinas Inc.	Buncombe	106,000	0.54
	Progress Energy Carolinas Inc.	Chatham	101,300	0.51
	Progress Energy Carolinas Inc.	Wayne	106,100	0.54
	Progress Energy Carolinas Inc.	Person	46,300	0.24
	Progress Energy Carolinas Inc.	New Hanover County	166,000	0.84
	Progress Energy Carolinas Inc.	Robeson	47,000	0.24
	Progress Energy Carolinas Inc.	Person	212,800	1.08
	Duke Energy Corp.	Gaston	143,400	0.73
	Duke Energy Corp.	Rowan	121,900	0.62
	Duke Energy Corp.	Cleveland	96,900	0.49
	Duke Energy Corp.	Rockingham	28,500	0.14
	Duke Energy Corp.	Catawba	33,500	0.17
	Duke Energy Corp.	Gaston	93,100	0.47
	Duke Energy Corp.	Stokes	41,400	0.21
North Dakota	Basin Electric Power Cooperative	Mercer	194,800	0.99
Ohio	Cardinal Operating Co.	Jefferson	490,400	2.49
	Cincinnati Gas & Electric Co.	Clermont	76,700	0.39
	Cincinnati Gas & Electric Co.	Hamilton	224,300	1.14
	Columbus Southern Power Co.	Coshocton	21,200	0.11
	Columbus Southern Power Co.	Pickaway	10,600	0.05
	Dayton Power & Light Co.	Adams	653,300	3.32
	Dayton Power & Light Co.	Adams	252,600	1.28
	Ohio Power Co.	Washington	143,400	0.73
	Ohio Power Co.	Gallia	90,700	0.46
	Ohio Valley Electric Corp.	Gallia	231,500	1.18
Oklahoma	Western Farmers Electric Cooperative Inc.	Choctaw	16,560	0.08
Pennsylvania	Pennsylvania Power Co.	Beaver	568,400	2.89
	PPL Corp.	Northampton	37,300	0.19

(*Continued*)

Table 5.13 List of Ash Ponds in the United States—*cont'd*

State	Facility Name	County	Capacity, Tons	Percent of Reported Capacity
	Sunbury Generation LLC	Snyder	500	0.00
South Carolina	Progress Energy Carolinas Inc.	Darlington	62,200	0.32
	Duke Energy Corp.	Anderson	63,500	0.32
	South Carolina Electric & Gas Co.	Colleton	101,100	0.51
	South Carolina Electric & Gas Co.	Aiken	12,500	0.06
	South Carolina Public Service Authority	Berkeley	10,900	0.06
	South Carolina Public Service Authority	Horry	7,000	0.04
	South Carolina Public Service Authority	Berkeley	34,900	0.18
	South Carolina Public Service Authority	Georgetown	8,950	0.05
Tennessee	Tennessee Valley Authority	Anderson	22,400	0.11
	Tennessee Valley Authority	Sumner	180,500	0.92
	Tennessee Valley Authority	Hawkins	10,000	0.05
	Tennessee Valley Authority	Humphreys	53,700	0.27
	Tennessee Valley Authority	Roane	325,900	1.65
Texas	Topaz Power Group LLC	Goliad	63,500	0.32
	Lower Colorado River Authority	Fayette	39,910	0.20
	Southwestern Electric Power Co.	Harrison	120,000	0.61
	TXU Electric Co.	Milam	314,400	1.60
Utah	Los Angeles City of	Millard	96,700	0.49
	Kennecott Utah Copper Corp.	Salt Lake	34,300	0.17
Virginia	Appalachian Power Co.	Giles	5,800	0.03
	Virginia Electric & Power Co.	Fluvanna	85,000	0.43
	Virginia Electric & Power Co.	Chesterfield	322,600	1.64
	Virginia Electric & Power Co.	Chesapeake	34,800	0.18
West Virginia	Appalachian Power Co.	Putnam	391,900	1.99
	Appalachian Power Co.	Kanawha	1,600	0.01
	Appalachian Power Co.	Mason	9,500	0.05
	Central Operating Co.	Mason	137,100	0.70
	Ohio Power Co.	Marshall	48,700	0.25
	Ohio Power Co.	Marshall	307,400	1.56
Wisconsin	Wisconsin Power & Light Co.	Columbia	11,000	0.06
Wyoming	Basin Electric Power Cooperative	Platte	79,100	0.40
	PacifiCorp.	Lincoln	119,000	0.60
	PacifiCorp.	Campbell	28,000	0.14

their age and integrity. Many are old and estimates are not entirely reliable or at least the basis for reported capacities not transparent. They represent potential risks to the public and a major management challenge for the utility industry.

Recommended Sources

Asai, Akihisu, et al. System Outline and Operational Status of Karita Power Station New Unit 1 (PFBC), *JSME International Journal*, Series B 47 (2) (2004) 193–199. Available at, http://www.jstage.jst.go.jp/article/jsmeb/47/2/193/_pdf.

Asia-Pacific Partnership on Clean Development and Climate (APP). Brown Coal Drying Technologies. APP Joint Meeting of Cleaner Fossil Energy and Power Generation and Transmission Task Forces. April 1, 2008. Available at http://www.asiapacificpartnership.org/pdf/CFE/meeting_melbourne/SurveyofBrownCoalDryingTechnologies-Godfrey.pdf.

Benson, Lewis, et al. Control of Sulfur Dioxide and Sulfur Trioxide Using By-Product of a Magnesium-Enhanced Lime FGD System. Presented at ICAC Forum '03 Multi-Pollutant Emission Controls & Strategies, Multi-Pollutant Emission Controls & Strategies, Nashville, TN, October 14–15, 2003. Available at, http://www.carmeusena.com/files/files/techpapersreports/carmeuse_icac_20forum_2003.pdf, October 14–15, 2003.

Benson, Lewis: New Magnesium-Enhanced Lime Flue Gas Desulfurization Process. Technical Paper Report, Carmeuse North America. Pittsburgh, PA. Available at http://www.carmeusena.com/files/files/TechPapersReports/fgd_new_magnesium.pdf

Biello David: Chemical Process Makes Fuel from Carbon Dioxide, Scientific American. (September 16, 2006). http://www.scientificamerican.com/article.cfm?id=chemical-process-makes-fu.

Canada's Integrated Carbon Dioxide (CO_2) Capture and Storage Initiative. ICO2N. http://www.ico2n.com.

Canadian CCS Projects. ICO2N. http://www.ico2n.com/ccs-in-canada/canadas-ccs-story/first-projects-in-canada.

Carbon Capture and Storage in Canada. Deloitte. http://www.deloitte.com/view/en_CA/ca/industries/energyandresources/1f769be5d3752210VgnVCM100000ba42f00aRCRD.htm.

Carbon Capture and Utilization Using CO_2 to Manufacture Fuel, Chemicals, and Materials. http://www.policyinnovations.org/ideas/policy_library/data/01612.

Casey Allen: Carbon Cemetery, Canadian Geographic Magazine (January/February 2008) 61.

Chang Kenneth, Scientists Would Turn Greenhouse Gas Into Gasoline, New York Times (February 19, 2008). http://www.nytimes.com/2008/02/19/science/19carb.html?_r=1.

Consumers Energy Company, Amendment to Application No. 341-07 Consumers Energy Company – Bay Count Top-Down BACT Analysis for IGCC, Consumers Energy Company, Jackson, MI, April 2008. Available at, http://www.deq.state.mi.us/aps/downloads/permits/CFPP/2007/341-07/Top-Down%20BACT%20Analysis%20for%20IGCC.pdf, April 2008.

Consumers Energy Company, Balanced Energy Initiative: Electric Generation Alternatives Analysis, Consumers Energy Company, Jackson, MI, June 2009. Available at, http://www.deq.state.mi.us/aps/downloads/permits/PubNotice/341-07/AlternativesAnalysis.pdf.

Davenport WG, et al. Sulfuric Acid Manufacture, Southern African Pyrometallurgy. South African Institute of Mining and Metallurgy, Johannesburg, March 5–8, 2006.

Demonstrating Carbon Capture and Storage in Canada. NRCan. 2010. http://www.nrcan-rncan.gc.ca/eneene/pubpub/pdf/ccscsc-eng.pdf.

Demonstration Project The Netherlands: Zero Emission Power Plant. http://www.createacceptance.net/fileadmin/create-acceptance/user/docs/Demonstration_project_The_Netherlands.pdf.

Enhance Energy Inc. | Enhanced Oil Recovery, Carbon Capture and Storage. Enhanceenergy.com. http://www.enhanceenergy.com/.

Escosa M. Jesús, Luis M. Romeo, Optimizing CO_2 Avoided Cost by Means of Repowering. Applied Energy 86 (2009) 2351–2358.

Foster Wheeler North America Corp, Utility CFB Goes Supercritical—Foster Wheeler's Lagisza 460 MWe Operating Experience and New 600–800 MWe Designs, Prepared by James Utt, Arto Hotta, and Stephen Goidich, Foster Wheeler North America Corp., Clinton, NJ, for presentation at Coal-Gen 2009, Charlotte, NC, August 19–21, 2009. Available at, http://www.fwc.com/publications/tech_papers/files/TP_CFB_09_12.pdf, August 19–21, 2009.

Graham, James. Microwaves for Coal Quality Improvement: The Drycol Project. DBAGlobal Australia, Milton Queensland, Australia. Presented at the SACPS/International Pittsburgh Coal Conference 2007, Johannesburg, South Africa, September 10–14, 2007. Available at http://www.drycol.com/downloads/Drycol%20Paper%20ACPS-1%20060608.pdf.

He, Boshu, et al. Temperature Impact on SO_2 Removal Efficiency by Ammonia Gas Scrubbing, Energy Conversion and Management 44 (2002) 2175–2188. http://www.sciencedirect.com/science?_ob=ArticleURL&_udi=B6V02-3WRC6DR-3&_user=775537&_origUdi=B6V2P-47CY482-G&_fmt=high&_coverDate=07%2F31%2F1998&_rdoc=1&_orig=article&_acct=C000042938&_version=1&_urlVersion=0&_userid=775537&md5=e2354e5058cba8f383722059ffecd1cd.

Hong BD, Slatick ER: Carbon Dioxide Emission Factors for Coal. Originally published in *U.S. Energy Information Administration, Quarterly Coal Report*, January–April 1994, DOE/EIA-0121(94/Q1). Washington, DC, August 1994: 1–8. Available at http://www.eia.doe.gov/cneaf/coal/quarterly/co2_article/co2.html.

IEA Clean Coal Centre. G8 Case Studies by the IEA Clean Coal Centre. Prepared by Colin Henderson, IEA Clean Coal Centre, London, United Kingdom for presentation at Third International Conference on Clean Coal Technologies for Our Future, Sotacarbo Coal Research Centre, Carbonia Sardinia, Italy. May 15–17, 2007. Available at http://www.ieacoal.org/publishor/system/component_view.asp?LogDocId=81704&PhyDocId=6360.

J-POWER (Electric Power Development Co., Ltd.). Annual Report 2009. Electric Power Development Co., Ltd., Tokyo, Japan. October 2009. Available at http://www.jpower.co.jp/english/ir/pdf/2009.pdf.

Kaplan, Ozge P, et al. Is It Better to Burn or Bury Waste for Clean Electricity Generation? Environmental Science & Technology 43 (6) (2009) 1711–1717.

Korhonen S, et al. Methane and Nitrous Oxide Emissions in the Finnish Energy Production, Fortum Tech (2001) 4615.

Lafleur Paul: Geochemical Soil Gas Survey, A Site Investigation of SW30-5-13-W2M, Weyburn Field, Saskatchewan (August 27, 2010).http://www.ecojustice.ca/media-centre/media-release-files/petro-find-geochem-ltd.-report/at_download/file.

Levasseur AA, et al. Alstom's Oxy-Firing Technology Development and Demonstration—Near Term CO_2 Solutions, Presented at the 34th International Technical Conference on Clean Coal & Fuel Systems, Clearwater, FL., May 31–June 4, 2009. Available at, http://www.netl.doe.gov/technologies/coalpower/ewr/co2/pubs/5290%20Alstom%20oxycombustion%20paper%20Clearwater%20jun09.pdf, May 31–June 4, 2009.

Maziuk John, John Kumm, Comparison of Dry Injection Acid-Gas Control Technologies, Presented at the 95th Annual Conference and Exhibition of the Air and Waste Management Association, Baltimore, MD, June 23–27, 2002.

Michigan Department of Natural Resources and Environment. PTI Application No. 341-07. December 29, 2009. Available at http://www.deq.state.mi.us/aps/downloads/permits/CFPP/2007/341-07/341-07.htm.

Mineral Carbonation and Industrial Uses of Carbon Dioxide. http://www.ipcc.ch/pdf/special-reports/srccs/srccs_chapter7.pdf.

National Energy Technology Laboratory (NETL). Cost and Performance Baseline for Fossil Energy Plants, Volume 1: Bituminous Coal and Natural Gas to Electricity, Revision 1. DOE/NETL-2007/1281. U.S. Department of Energy, National Energy Technology Laboratory, Pittsburgh, PA. August 2007. Available at http://www.netl.doe.gov/energyanalyses/pubs/Bituminous%20Baseline_Final%20Report.pdf.

National Energy Technology Laboratory (NETL), Reducing CO_2 Emissions by Improving the Efficiency of the Existing Coal-Fired Power Plant Fleet. DOE/NETL-2008/1329, U.S. Department of Energy, National Energy Technology Laboratory, Pittsburgh, PA, July 23, 2008. Available at, http://www.netl.doe.gov/energy-analyses/pubs/CFPP%20Efficiency-FINAL.pdf, July 23, 2008.

National Energy Technology Laboratory (NETL). Overview of DOE's Gasification Program. Presentation by Jenny Tennant, Technology Manager, Gasification, U.S. Department of Energy, National Energy Technology Laboratory, Pittsburgh, PA. January 25, 2010. Available at http://www.netl.doe.gov/technologies/

coalpower/gasification/pubs/pdf/DOE%20Gasification%20Program %20Overview%202010%2001-25%20v1v.pdf.

National Energy Technology Laboratory (NETL). CCPI/Clean Coal Demonstrations Nucla CFB Demonstration Project, Project Fact Sheet. U.S. Department of Energy, National Energy Technology Laboratory, Pittsburgh, PA. Accessed June 21, 2010. Available at http://www.netl.doe.gov/technologies/coalpower/cctc/summaries/nucla/nuclademo.html.

National Energy Technology Laboratory (NETL). CCPI/Clean Coal Demonstrations Tidd PFBC Demonstration Project, Project Fact Sheet. U.S. Department of Energy, National Energy Technology Laboratory, Pittsburgh, PA. Accessed June 21, 2010. Available at http://www.google.com/imgres?imgurl=http://www.netl.doe.gov/technologies/coalpower/cctc/summaries/tidd/images/tidd_plant_bw.jpg&imgrefurl=http://www.netl.doe.gov/technologies/coalpower/cctc/summaries/tidd/tidddemo.html&usg=__PMP6wAdGSJtmGws-CiNKKCwryUc=&h=

NETL Carbon Sequestration NETL website. http://www.netl.doe.gov/technologies/carbon_seq/infrastructure/knowledgesharing.html.

Norway: StatoilHydro's Sleipner Carbon Capture and Storage Project Proceeding Successfully. Energy-pedia. March 8, 2009. http://www.energy-pedia.com/article.aspx?articleid=134204.

Pacific Northwest National Laboratory (PNNL), An Assessment of the Commercial Availability of Carbon Dioxide Capture and Storage Technologies as of June 2009. PNNL-18520, Pacific Northwest National Laboratory, Richland, WA, June 2009, http://www.pnl.gov/main/publications/external/technical_reports/PNNL-18520.pdf.

Quest Project Canada. Shell.ca. October 8, 2009. http://www.shell.ca/home/content/can-en/aboutshell/our_business/business_in_canada/upstream/oil_sands/quest/.

Ravi K. Srivastava, Jozewicz Wojciech, Flue Gas Desulfurization: The State of the Art, Journal of the Air & Waste Management Association 51 (2001) 1676–1688.

Ravi K. Srivastava, Jozewicz Wojciech, Singer Carl, SO_2 Scrubbing Technologies: A Review, Environmental Progress 20 (4) (2001) 219–228.

Retzlaff, Klaus M, Anthony Ruegger W: Steam Turbines for Ultrasupercritical Power Plants. GE Power Generation, GER-3945A. Available at http://www.gepower.com/prod_serv/products/tech_docs/en/downloads/ger3945a.pdf.

Robert Peltier, Genesee Phase 3, Edmonton, Alberta, Canada, Power (July/August 2005). Available at http://www.epcor.ca/SiteCollectionDocuments/Corporate/pdfs/publications%20and%20newsletters/plattspower05.pdf.

Robert Peltier, Plant Efficiency: Begin with the Right Definitions, Power (February 1, 2010). Available at http://www.powermag.com/gas/Plant-Efficiency-Begin-with-the-Right-Definitions_2432.html.

RWE Power. Niederaussem Power Plant: A Plant Full of Energy. RWE Power Aktiengesellschaf, Essen, Germany. October 2004. Available at http://www.debriv.de/tools/download.php?filedata=1218532199.pdf&filename=Kraftwerk%20Niederaussem%20(englisch).pdf&mimetype=application/pdf.

Sargent, Lundy LLC: Coal-Fired Power Plant Heat Rate Reductions. Report Number SL-009597, Sargent & Lundy, Chicago, IL, January 22, 2009. Available at http://www.epa.gov/airmarkt/resource/docs/coalfired.pdf.

Schimmoller Brian: Avista Kettle Falls, Power Engineering. (October 2003). Available at http://www.powergenworldwide.com/index/display/articledisplay/189185/articles/powerengineering/volume-107/issue-10/departments/managing-the-plant/avista-kettle-falls.html.

Shand Mark A: A New Look at SO_2 Removal. Available at http://www.risiinfo.com/technologyarchives/chemicalsnews, April 28, 2009.

Stenzel, William C, et al: Repowering Existing Fossil Steam Plants. December 11, 1997. Available at http://soapp.epri.com/press/default.htm.

U.S. Department of Energy (DOE). Clean Coal Technology Power Plant Optimization Demonstration Projects Topical Report Number 25. September 2007. Available at http://www.netl.doe.gov/technologies/coalpower/cctc/topicalreports/pdfs/topical25.pdf

U.S. Department of Energy (DOE). Electric Utility Engineer's FGD Manual, Vol. 1—FGD Process Design. U.S. Department of Energy, Office of Fossil Energy, Morgantown, WV. March 1996. Available at http://www.netl.doe.gov/technologies/coalpower/ewr/pubs/fgdmanual_vol1.pdf.

U.S. Department of Energy (DOE), Innovative Drying Technology Extracts More Energy from High Moisture Coal, Fossil Energy Techline (March 11, 2010). Available at http://fossil.energy.gov/news/techlines/2010/10006-CCPI_Technology_Goes_Commercial.html.

U.S. Department of Energy (DOE). Technical Workshop Report: Improving the Efficiency of Coal-fired Power Plants in the United States. Workshop sponsored by the U.S. Department of Energy and National Energy Technology Laboratory. Baltimore, MD. February 24–25, 2010. Available at http://www.netl.doe.gov/energyanalyses/refshelf/detail.asp?pubID=306.

U.S. Department of Energy (DOE). Technical Workshop Report: Opportunities to Improve the Efficiency of Existing Coal-Fired Power Plants. Workshop sponsored by the U.S. Department of Energy and National Energy Technology Laboratory. Rosemont, IL. July 15–16, 2009. Available at http://www.netl.doe.gov/energyanalyses/pubs/NETL%20Power%20Plant%20Efficiency%20Workshop%20Report%20Final.pdf.

U.S. Energy Information Administration (U.S. EIA). EIA-923 (Schedule 2)—Monthly Utility and Nonutility Fuel Receipts and Fuel Quality Data (2008 Final Data). U.S. Energy Information Administration, Office of Coal, Nuclear, Electric and Alternate Fuels, U.S. Department of Energy, Washington, DC 20585. Accessed June 21, 2010. Available at http://www.eia.doe.gov/cneaf/electricity/page/eia423.html.

U.S. Energy Information Administration (U.S. EIA). Electric Power Annual 2008. DOE/EIA-0348(2008). U.S. Energy Information Administration, Office of Coal, Nuclear, Electric and Alternate Fuels, U.S. Department of Energy, Washington, DC 20585. January 2010. Available at http://www.eia.doe.gov/cneaf/electricity/epa/epa_sum.html.

U.S. Environmental Protection Agency (EPA), Database of information collected in the Electric Utility Steam Generating Unit Mercury Emissions Information Collection Effort. OMB Control No. 2060-0396, U.S. Environmental Protection Agency, Office of Air Quality Planning and Standards, Research Triangle Park, NC, April 2001.

U.S. Environmental Protection Agency (EPA), Environmental Footprints and Costs of Coat-Based Integrated Gasification Combined Cycle and Pulverized Coal Technologies. EPA. EPA-600/R-06/006, U.S. Environmental Protection Agency, National Risk Management Research Laboratory, Research Triangle Park, NC, July 2006. Available at http://www.epa.gov/oar/caaac/coaltech/2007_01_epaigcc.pdf.

U.S. Environmental Protection Agency (EPA), Multipollutant Emission Control Technology Options for Coal-Fired Power Plants. EPA. EPA-600/R-05/034, U.S. Environmental Protection Agency, National Risk Management Research Laboratory, Research Triangle Park, NC, March 2005. Available at, http://www.epa.gov/airmarkt/resource/docs/multipreport2005.pdf.

Utility Air Regulatory Group (UARG), A Review of Carbon Capture and Sequestration (CCS) Technology, December 2008.

Vattenfall, Nordjylland Power Station: The World's Most Efficient Coal-Fired CHP Plant, Vattenfall, A/S Copenhagen, Denmark (2006). Available at http://www.vattenfall.dk/da/file/nordjyllanduk080121_7841588.pdf.

VGB Powertech and Evonik Energy Services. Supercritical and Ultra Supercritical Technology. In Power Plant Performance Reporting and Improvement under the Provision of the Indian Energy Conservation Act: Output 1.1: Best practice performance monitoring, analysis of performance procedures, software and analytical tools, measuring instrumentation, guidelines or best practice manuals and newest trends. (Annexure XIV). Indo German Energy Programme. 2008. Retrieved from http://www.emtindia.net/PowerPlantComponent/Output1.1/Output1.1.pdf.

Wagner Leonard: Carbon Capture and Storage, Moraassociates.com (2007).http://www.moraassociates.com/publications/0701%20Carbon%20capture%20and%20storage.pdf.

Weijuan Yang, et al. Nitrous Oxide Formation and Emission in Selective Non-Catalytic Reduction, Energy Power Engineering China (2007) 228–232.

Glossary

ASTM ASTM International, as a standards development body, provides technical standards, which include metals, petroleum, construction, environment, and many other related standards, test methods, specifications, and guides.

Accessed Coal deposits that have been prepared for mining by construction of portals, shafts, slopes, drifts, and haulage ways; by removal of overburden; or by partial mining (see also *virgin coal*).

Accessibility Factor The estimated regional ratio of accessible reserve base to the demonstrated reserve base or of accessible resources to identified resources.

Accessibility In reference to coal resources (core meaning), the absence of land-use restrictions and the assumption that ownership or leaseholds will be obtainable for mining (see also *environmental restrictions, industrial restrictions*). Many technological restrictions were traditionally applied as demonstrated reserve base criteria, but (extended meaning) with the advent of available resource studies, specific technologic restrictions may be incorporated in accessibility factors (see also *restricted resources*).

Accessible Reserve Base The portion of the demonstrated reserve base estimated by EIA to be accessible, determined by application of one or more accessibility factors within an area. An accessible reserve base may be referred to as accessible resources because it is a subset of accessible resources and is usually part of a single resource study.

Accessible Resources The portion of identified resources estimated to be accessible, determined by application of one or more accessibility factors within an area.

Admixtures Material other than water, aggregates, hydraulic cement, and fiber reinforcement used as an ingredient of concrete or mortar and added to the batch immediately before or during its mixing. Fly ash is used as an admixture in concrete.

Anthracite Coal A type of coal that has the highest carbon content and the lowest moisture and ash content. Anthracite burns slowly and makes a good heating fuel for homes. The United States has about 7.3 billion tons of anthracite, most of which can be found in Pennsylvania.

Ash Hopper A collection chamber, usually in the shape of an inverted tapered cone or pyramid, where collected ash is temporarily stored. As the ash is collected from the flue gas, it drops into the large end of the ash hopper and accumulates. The small end of the hopper is closed with a valve that is periodically opened to remove ash when the hopper is full.

Ash Impoundment Pond An impoundment or surface impoundment used to store or dispose of ash primarily from the combustion of coal. A type of waste management facility consisting of an excavated, a dammed, or a diked reservoir in which coal ashes are stored for future removal or disposed of as a slurry or sludge. The coal ash solids settle out and leave relatively clear water at the surface that is discharged through a designed and managed outlet structure to a nearby stream, surface water, or plant process water system. Ash pond designs reflect local site conditions; federal and state regulations; and whether fly ash, bottom ash, boiler slag, or a combination of coal ashes is disposed in the ash pond. While some electric utility generating power companies combine the ashes during storage or disposal, other power companies use separate ash ponds for fly ash, bottom ash, and boiler slag. The ash pond is referred to as a bottom ash pond, fly ash pond, or boiler slag pond when it receives one type of ash. Also a large ash pond is referred to as an ash impoundment, ash reservoir, or surface impoundment.

As-Received Condition or As-Received Basis Represents an analysis of a sample as received at a laboratory.

Availability In reference to coal resources, the absence of land-use or environmental restrictions and technological restrictions.

Available Reserves In EIA coal supply modeling, the difference between estimated recoverable reserves and recoverable reserves at active mines; in modeling context, these reserves are considered not presently obligated for existing mines and, therefore, would be available to supply new mines in the future.

Available Resources In U.S. Geological Survey studies, the quantity of remaining identified resources available for development and potential extraction at the time of determination after adjusting for geologic considerations, land-use restrictions, and/or technological restrictions (see also *accessible reserve base*).

Bed, Coalbed All the coal and partings lying between a roof and floor.

Bench Sample A face or channel sample taken of just that contiguous portion of a coalbed that is considered practical to mine, also known as a "bench." For example, bench samples may be taken of minable coal where impure coal that makes up part of the geologic coalbed is likely to be left in the mine, or where thick partings split the coal into two or more distinct minable seams, or where extremely thick coalbeds cannot be recovered by normal mining equipment so that the coal is mined in multiple passes, or benches, usually defined along natural bedding planes.

Bench A subdivision and (or) layer of a coal bed separated from other layers by partings of non-coal rock.

Beneficiation Improvement of the chemical or physical properties of a raw material or intermediate product by the removal or modification of undesirable components or impurities. The removal of unburned carbon in fly ash is an example of beneficiation of the raw fly ash.

Bituminous Coal The most common type of coal in the U.S. "Bituminous" is a term that refers to the "rank" of the coal, which is classified based on its content of fixed carbon, volatile matter, and heating value. There are several groups of bituminous coal: high-volatile A, B, and C; medium-volatile; and low-volatile bituminous. Bituminous coal generally has a high heating value and low moisture content. It is mainly burned to produce electricity but is also used to produce metallurgical coke.

Byproduct Something produced as a secondary result of the manufacture or production of something else.

Calcined Heating of a substance to a temperature until it decomposes.

Carbonaceous Containing or yielding coal or carbon, an energy source.

Carbon Flotation The flotation wet process consists of raw feed silos, wet conditioning systems, processing and secondary flotation cells, dewatering equipment, drying equipment, and finished product silos. Normally, a frothing agent is used to coat the unburned carbon-forming hydrophobic carbon materials. Air is introduced into the bottom of the cells, and the air carries the hydrophobic unburned carbon to the top, where it is skimmed off, dried, and can be introduced into the furnace as fuel. The remaining fly ash is dewatered and dried, ready for shipment to a concrete plant.

Cement Finely powdered mixtures of inorganic compounds that when combined with water will set and harden.

Char A combustible residue of carbonaceous feedstock.

Class C Fly Ash Fly ash that meets criteria defined in ASTM C 618 for use in concrete.

Class F Fly Ash Fly ash that meets criteria defined in ASTM C 618 for use in concrete.

Clean Coal Yield In EIA models, the portion of reserves that will be salable coal after preparation, if any is done (see also *salable coal*).

Coal A brown to black combustible sedimentary rock (in the geologic sense) composed principally of consolidated and chemically altered plant remains; solid hydrocarbon fuel formed by ancient decomposition of woody substance under conditions of heat and pressure. All solid fuels are classified as anthracite, bituminous, subbituminous, or lignite by ASTM and in ASTM D388-77.

Coal Preparation/Washing The treatment of coal to reject waste. In its broadest sense, preparation is any processing of mined coal to prepare it for market, including crushing and screening or sieving the coal to reach a uniform size, which normally results in removal of some non-coal material. The term "coal preparation" most commonly refers to processing, including crushing and screening, passing the material through one or more processes to remove impurities, sizing the product, and loading for shipment. Many of the processes separate rock, clay, and other minerals from coal in a liquid medium; hence, the term "washing" is widely used. In some cases coal passes through a drying step before loading.

Coal Sampling The collection and proper storage and handling of a relatively small quantity of coal for laboratory analysis. Sampling may be done for a wide range of purposes, such as coal resource exploration and assessment, characterization of the reserves or production of a mine, to characterize the results of coal cleaning processes, to monitor coal shipments or receipts for adherence to coal quality contract specifications, or to subject a coal to specific combustion or reactivity tests related to the customer's intended use. During predevelopment phases, such as exploration and resource assessment, sampling typically is from natural outcrops, test pits, old or existing mines in the region, drill cuttings, or drilled cores. Characterization of a mine's reserves or production may use sample collection in the mine, representative cuts from coal conveyors or from handling and loading equipment, or directly from stockpiles or shipments (coal rail cars or barges). Contract specifications rely on sampling from the production flow at the mining or coal-handling facility or at the loadout, or from the incoming shipments at the receiver's facility. In all cases, the value of a sample taken depends on its being representative of the coal under consideration, which in turn requires that appropriate sampling procedures be carefully followed.

Coal Supply Region An area in which the EIA coal reserves data are aggregated and allocated to a set of uniform, typical criteria for purposes of modeling. The criteria of a coal supply region may include coal heat and sulfur content or other quality parameters, coal rank, geographic continuity, traditional mining regions, state or county boundaries, transportation corridors and barriers, and marketing factors. Coal supply regions may vary for different modeling criteria; they may include the coal reserves of an entire state or a contiguous group of states; some major producing states may be split into more than one region.

Coalbed All the coal and partings lying between a roof and floor.

Coal-Producing Region An area that collectively encompasses a group of geographically contiguous or logically associated states or areas that currently or historically mine and market coal.

Coke A hard, porous material produced by heating bituminous coal to a high temperature in the absence of air. It is used in the steel industry to melt and reduce iron ore to metallic iron, which is the main ingredient in steel (metallurgical coke).

Column Sample A channel or drill core sample taken to represent the entire geologic coalbed; it includes all partings and impurities that may exist in the coalbed.

Combustion A chemical process in which oxygen reacts strongly with other substances to produce a flame (heat and light).

Committed Reserves In EIA coal supply modeling, synonymous with recoverable reserves at active mines; in modeling context, these reserves are considered obligated for existing mines and, therefore, not part of the reserves that would be available to supply new mines in the future.

Compliance Coal A coal or a blend of coals that meet sulfur dioxide emission standards for air quality without the need for flue-gas desulfurization.

Composite Sample A recombined coalbed sample produced by averaging together thickness-weighted coal analyses from partial samples of the coalbed, such as from one or more bench samples, from one or more mine exposures or outcrops where the entire bed could not be accessed in one sample, or from multiple drill cores that were required to retrieve all local sections of a coal seam.

Concrete A hard compact building material formed when a mixture of cement, sand, gravel, and water undergoes hydration.

Cumulative Depletion The sum in tons of coal extracted and lost in mining to a stated date for a specified area or a specified coal bed.

Cyclone Boiler A type of coal-fired boiler. The coarsely pulverized coal undergoes slagging combustion in a cylindrical (cyclone) burner. Some wet-bottom boilers are not cyclone-fired. The primary byproduct is a glassy slag referred to as boiler slag, which is in great demand for beneficial use, but the supplies are declining.

Demonstrated Reserve Base (DRB) A collective term for the sum of coal in both measured and indicated resource categories of reliability; the DRB represents 100% of coal in place as of a certain date. Includes beds of bituminous coal and anthracite 28 inches or more thick and beds of subbituminous coal 60 inches or more thick that can be surface mined. Includes also thinner and/or deeper beds that presently are being mined or for which there is evidence that they could be mined commercially at this time. Represents that portion of the identified resources of coal from which reserves are calculated.

Demonstrated Resources Same qualifications as identified resources, but includes measured and indicated degrees of geologic assurance and excludes the inferred.

Depleted Resources Resources that have been mined; includes coal recovered, coal lost in mining, and coal rendered subeconomic as a result of mining the recovered coal. (See also *cumulative depletion*.)

Depletion Factor The multiplier of the tonnage produced that takes into account both the tonnage recovered and the tonnage lost due to mining. The depletion factor is the reciprocal of the recovery factor in relation to a given quantity of production.

Depletion The subtraction of both the tonnage produced and the tonnage lost in mining from the demonstrated reserve base and identified resources to determine the remaining tonnage as of a certain time.

Dry, Mineral-Matter-Free Basis A type of calculated analytical value of a coal sample expressed as if the total moisture and mineral matter had been removed. Mineral matter free is not the same as ash free.

Economic Term that implies that profitable extraction or production under realistic investment assumptions has been established, analytically demonstrated, or assumed with reasonable certainty.

Electrostatic Precipitator (ESP) Hardware that removes fly ash from the flue gas by producing an electric charge on the fly ash and collecting it electrostatically.

Endothermic Absorbing or requiring heat to maintain chemical reaction temperatures.

Environmental Restrictions Land-use or subsurface restrictions that constrain, postpone, or prohibit mining in order to protect environmental resources of an area; for example, surface or groundwater quality, air quality affected by mining, or plants or animals or their habitats.

Estimate A determination as to the amount or tonnage of coal in an area. The term "estimate" indicates that the quantities of resources are known imprecisely. An estimate differs from an assessment, which is an analysis of all data concerning an area's coal resources and reserves with the objective of reaching a judgment about the geologic nature and economic potential of the coal resources and reserves of the area.

Estimated Recoverable Reserves See *recoverable reserves.*

Exothermic Releasing or giving off heat during a chemical reaction.

Face Channel or Channel Sample A sample taken at the exposed coal in a mine by cutting away any loose or weathered coal and then collecting on a clean surface a sample of the coal seam by chopping out a channel of uniform width and depth; a face channel or face sample is taken at or near the working face, the most freshly exposed coal where actual removal and loading of mined coal is taking place. Any partings greater than $^3/_8$ inch and/or mineral concretions greater than ½ inch thick and 2 inches in maximum diameter are normally discarded from a channel sample so as better to represent coal that has been mined, crushed, and screened to remove at least gross non-coal materials.

Feedstock A raw material used by industry in the manufacture of a product.

Floor The upper surface of the stratum underlying a coal seam. In coals that were formed in persistent swamp environments, the floor is typically a bed of clay, known as "underclay," representing the soil in which the trees or other coal-forming swamp vegetation was rooted.

Flowable Fill Also termed "controlled low-strength material" or "CLSM," flowable fills are mainly used for trench filling, but may also be used in other applications such as backfilling bridge abutments and retaining walls. Flowable fill contains much less cement and higher proportions of fly ash and water than concrete. Sand is generally the only aggregate used. The fills are designed to have relatively low strength so that they can be excavated if necessary.

Flue-Gas Desulfurization (FGD) Removal of gaseous sulfur dioxide from boiler exhaust gas. Primary types of FGD processes are wet scrubbers, dry scrubbers, and sorbent injection. Sorbents include lime, limestone, sodium-based compounds, and high-calcium coal fly ash.

Gasification Conversion of low-value carboniferous feedstocks to higher-value gaseous fuels.

Gasifier Main component of a gasification system; it converts a solid or liquid into a gas by means of heat and pressure in the presence of oxygen.

Geologic Assurance State of sureness, confidence, or certainty of the existence of a quantity of resources based on the distance from points where coal is measured or sampled and on the abundance and quality of geologic data as related to thickness of overburden, rank, quality, thickness of coal, areal extent, geologic history, structure, and correlations of coal beds and enclosing rocks. The degree of assurance increases as the nearness to points of control, abundance, and quality of geologic data increases.

Geologic Considerations Conditions in the coal deposit or in the rocks in which it occurs that may complicate or preclude mining. Geologic considerations are evaluated in the context of the current state of technology and regulations, so the impact on mining may change with time.

Grade (of Coal) See *quality*.

Hydrocarbon A fuel composed of hydrogen and carbon.

Hypothetical Resources Undiscovered coal resources in beds that may reasonably be expected to exist in known mining districts under known geologic conditions. In general, hypothetical resources are in broad areas of coalfields where points of observation are absent and evidence is from distant outcrops, drill holes, or wells. Exploration that confirms their existence and better defines their quantity and quality would permit their reclassification as identified resources. Quantitative estimates are based on a broad knowledge of the geologic character of coalbed or region. Measurements of coal thickness are more than 6 miles apart. The assumption of continuity of coalbed is supported only by geologic evidence.

Identified Resources Specific bodies of coal whose location, rank, quality, and quantity are known from geologic evidence supported by engineering measurements. Included are beds of bituminous coal and anthracite 14 inches or more thick and beds of subbituminous coal and lignite 30 inches or more thick that occur at depths to 6,000 feet and whose existence and quantity have been delineated within specified degrees of geologic assurance as measured, indicated, or inferred.

Indicated Resources Coal for which estimates of the rank, quality, and quantity have been computed partly from sample analyses and measurements and partly from reasonable geologic projections. Indicated resources are computed partly from specified measurements and partly from projection of visible data for a reasonable distance on the basis of geologic evidence. The points of observation are 0.5 to 1.5 miles apart. Indicated coal is projected to extend as a 0.5-mile-wide belt that lies more than 0.25 miles from the outcrop or points of observation or measurement.

Industrial Restrictions Land-use restrictions that constrain, postpone, or prohibit mining in order to meet other industrial needs or goals; for example, resources not mined due to safety concerns or due to industrial or societal priorities, such as to preserve oil or gas wells that penetrate the coal reserves; to protect surface features such as pipelines, power lines, or company facilities; or to preserve public or private assets, such as highways, railroads, parks, or buildings.

Inferred Reserve Base The resources in the inferred reliability category that meet the same criteria of bed thickness and depth from surface as the demonstrated reserve base.

Inferred Resources Coal in unexplored extensions of demonstrated resources for which estimates of the quality and size are based on geologic evidence and projection. Quantitative estimates are based largely on broad knowledge of the geologic character of the bed or region and where few measurements of bed thickness are available. The estimates are based primarily on an assumed continuation from demonstrated coal for which there is geologic evidence. The points of observation are 1.5 to 6 miles apart. Inferred coal is projected to extend as a 2.25-mile-wide belt that lies more than 0.75 miles from the outcrop or points of observation or measurement.

Inorganic Composed of minerals other than living materials or carbon.

Isopach A line on a map drawn through points of equal thickness of a designated unit (such as a coal bed).

Landfill A disposal facility where waste is placed in or on land; a facility where "dry" (actually moistened) coal combustion or flue-gas cleaning byproducts (CCBs) are placed for disposal in or on land. CCBs are transported to this facility directly from the coal-fired plant after they are produced or after they are dredged from storage impoundments that are used as interim facilities. The disposed CCBs remain in the landfill after closure. Also these CCBs are dry (moistened) and have the consistency of soil. As a result, dikes are not required to provide stability. Most large landfills are divided into sections or cells, and the CCBs are placed in layers that are referred to as lifts that can vary in thickness. Typically, captive CCB landfills are designed and permitted to receive only CCBs and are classified as mono-fills.

Land-Use Restrictions Constraints placed upon mining by societal policies to protect surface features or entities that could be affected by mining. Because laws and regulations may be modified or repealed, the restrictions, including industrial and environmental restrictions, are subject to change.

Lignite This is the lowest rank coal in the United States and thus has a low heating value and a high moisture content. If the lignite is of a sufficiently low rank, it might be termed "brown coal." Lignite is fairly soft and can be brown or black in color. It is burned to generate electricity.

Low-NOx Burners A component of coal-fired boilers that are used to generate electricity. These burners are designed to lower the combustion temperature to generate lower NOx emissions.

Marginal Reserves A term that implies the reserves border on being economic. See also *economic*.

Measured Resources Coal for which estimates of the rank, quality, and quantity have been computed, within a high degree of geologic assurance, from sample analyses and measurements from closely spaced and geologically well-known sample sites. Measured resources are computed from dimensions revealed in outcrops, trenches, mine workings, and drill holes. The points of observation and measurement are so closely spaced and the thickness and extent of coals are so well defined that (for older estimates) the tonnage was judged to be accurate within 20% of true tonnage (statistical measures of error are no longer considered reliable for most measured resources). Although the spacing of the points of observation necessary to demonstrate continuity of the coal differs from region to region according to the character of the coalbeds, the points of observation are not greater than 0.5 miles apart. Measured coal is projected to extend as a 0.25-mile-wide belt from the outcrop or points of observation or measurement.

Methane The primary constituent of natural gas.

Minable Capable of being mined under current mining technology and environmental and legal restrictions, rules, and regulations.

Organic Composed of living materials or carbon.

Original (Resources/Reserves) The amount of coal in the ground before any production.

Overburden Any material, consolidated or unconsolidated, that lies between a coal deposit and the surface. Overburden is reported in feet and (or) meters and used to classify the depth to an underlying coal bed.

Oxidation A chemical reaction in which oxygen is added to an element or compound.

Pet Coke Petroleum coke, a solid residue composed mainly of carbon that is left after volatile elements have been driven from coal or other petroleum material.

Pozzolan Primarily siliceous or siliceous and aluminous materials that will, in finely divided form and in the presence of moisture, chemically react with calcium oxide at ordinary temperatures to form compounds possessing cementitious properties.

Pozzolanic (activity/reactivity) The phenomenon of strength development that occurs when lime and certain aluminosilicates react at ambient temperatures in the presence of water.

Preparation Plant Broadly speaking, any facility where coal is prepared for market; the usual accepted meaning is a rather elaborate collection of facilities where coal is separated from its impurities, washed and sized, and loaded for shipment. Also known as a wash plant or coal washer.

Pulverized Coal Combustion (PCC) Any combustion process that uses very finely ground (pulverized) coal in the process. Pulverized coal combustion processes usually result in the production of bottom and fly ashes.

Quality or Grade An informal classification of coal relating to its suitability for use for a particular purpose. Refers to individual measurements such as heat value; fixed carbon; moisture; ash; sulfur; phosphorus; major, minor, and trace elements; coking properties; petrologic properties; and particular organic constituents. The individual quality elements may be aggregated in various ways to classify coal for such special purposes as metallurgical, gas, petrochemical, and blending usages.

Rank The classification of coal relative to other coals, according to their degree of metamorphism, or progressive alteration, in the natural series from lignite to anthracite (Standard Classification of Coal by Rank, 1992, American Society for Testing and Materials, ASTM Designation D-388-91a).

Recoverability In reference to accessible coal resources, the condition of being physically, technologically, and economically minable. Recovery rates and recovery factors may be determined or estimated for coal resources without certain knowledge of their economic minability; therefore, the availability of recovery rates or factors does not predict recoverability.

Recoverable Coal Coal that is, or can be, extracted from a coal bed during mining.

Recoverable Reserves at Active Mines The amount of *in situ* coal that can be recovered by mining existing reserves at mines reporting on Form EIA-7A (see also *committed reserves*).

Recoverable Reserves–Estimated Recoverable Reserves Reserve estimates (broad meaning) based on a demonstrated reserve base adjusted for assumed accessibility factors and recovery factors. EIA uses the term to distinguish estimated recoverable reserves, which are derived without specific economic feasibility criteria by factoring (downward) from a demonstrated reserve base for one or more study areas or regions, from recoverable reserves at active mines, which are aggregated (upward) from reserve estimates reported by currently active, economically viable mines on Form EIA-7A.

Recovery Factor The percentage of total tons of coal estimated to be recoverable from a given area in relation to the total tonnage estimated to be in the demonstrated reserve base. For the purpose of calculating depletion factors only, the estimated recovery factors for the demonstrated reserve base generally are 50% for underground mining methods and 80% for surface mining methods. More precise recovery factors can be computed by determining the total coal in place and the total coal recoverable in any specific locale.

Recovery Percentage/Rate The ratio of coal extracted from a bed as compared to the total quantity of coal originally in the bed.

Regional Reserves, Regional Reserve Estimates Same as reserves; EIA uses alternative wording to distinguish regional reserves, which are derived by factoring (downward) from a demonstrated reserve base for one or more study areas or regions, from reserves at active mines, which are aggregated (upward) from reserve estimates reported by individual mines on Form EIA-7A.

Remaining (Resources/Reserves) The amount of coal in the ground after some mining, excluding coal in the ground spoiled or left in place for which later recovery is not feasible.

Reserves at Active Mines The amount of *in situ* coal that can be recovered by mining existing reserves at mines reporting on Form EIA-7A (see also *committed reserves*).

Resources Naturally occurring concentrations or deposits of coal in the earth's crust, in such forms and amounts that economic extraction is currently or potentially feasible.

Restricted Coal Resources In U.S. Geological Survey studies, the quantity of remaining resources that are not available for development at the time of determination because of geologic considerations, land-use restrictions, and/or technological restrictions.

Restricted Resources Those parts of any resource category that are restricted or prohibited from extraction by laws or regulations. Also, coal or a portion of the coal in categorically minable depths or thicknesses that is not economic at the time of determination.

Roof The rock immediately above a coal seam. The roof is commonly a shale, often carbonaceous and softer than rocks higher up in the roof strata.

Run-of-Mine The raw coal recovered from a mine, prior to any treatment.

Salable Coal The shippable product of a coal mine or preparation plant. Depending on customer specifications, salable coal may be run-of-mine, crushed-and-screened (sized) coal, or the clean coal yield from a preparation plant.

Sample A representative fraction of a coal bed collected by approved methods, guarded against contamination or adulteration, and analyzed to determine the nature; chemical, mineralogic, and (or) petrographic composition; percentage or parts-per-million content of specified constituents; heat value; and possibly the reactivity of the coal or its constituents.

Scanning Electron Microscope (SEM) An electron-beam-based microscope that uses electrons to produce high-resolution three-dimensional images of specimens and allows magnifications of hundreds of times greater than a light microscope. An SEM scan is taken by scanning a focused probe across the surface of the sample to be studied. It can focus on both the "hills" and the "valleys" of an object at the same time so that sharp images of very fine details are produced. The more electrons a particular region emits, the brighter the image at that point. SEM images typically contain a good deal of topographical detail.

Scrubbers See *flue-gas desulfurization.*

Seam A bed of coal lying between a roof and floor. Equivalent term to "bed," commonly used by industry.

SEM See *scanning electron microscope.*

Slag Mineral waste left after coal has been gasified.

Slag Tap A term used for a coal-fired boiler wherein the ash is kept in a molten state and is removed, or "tapped," from the bottom of the boiler. The molten slag is quenched in water to form a glassy solid. Slag tapping is also done in blast furnaces. In this case, molten slag floats on top of the molten iron and is tapped from the surface of the iron.

Slurry A liquid mixture of water and an insoluble solid, such as coal.

Sorbent The material within a scrubber that is used to remove sulfur oxides (SOx) from flue gas during coal combustion. In most systems the sorbent is limestone or lime mixed with water.

Speculative Resources Undiscovered coal in beds that may occur either in known types of deposits in a favorable geologic setting where no discoveries have been made or in deposits that remain to be recognized. Exploration that confirms their existence and better defines their quantity and quality would permit their reclassification as identified resources.

Strip or Stripping Ratio The amount of overburden that must be removed to gain access to a unit amount of coal.

Strip or Surface Mining The extraction of coal by using surface mining methods such as area strip mining, contour strip mining, or open-pit mining. The overburden covering the coal is removed and the coal extracted using power shovels, front-end loaders, or similar heavy equipment.

Subbituminous Coal As the name implies, coal that is intermediate in rank between bituminous coal and lignite. It thus has an intermediate heating value, volatile matter content, and moisture content. It is burned to generate electricity.

Syngas A gaseous fuel derived (or synthesized) from a carbonaceous source such as coal and used as a substitute for a petroleum product.

Technological Restrictions Constraints related to economics and safety placed upon mining by contemporary technology or prescribed by law; the restrictions may change with advances in science or modifications in the law. For purposes of assessing impacts on minability, geologic considerations are included as technological restrictions. A secondary basis for accessibility is the technological restrictions that may affect economic minability of specific coal resources in a locality at the time of the evaluation. Technological restrictions include constraints on the economic or safe mining of the coal with contemporary technologies; these constraints are related to the nature of the coalbeds or local geology—for example, specific coalbed thickness or overburden characteristics known to deter economic mining of coal meeting broad regional DRB criteria, localized geologic structural problems, or unsafe or illegal proximity to another mine.

Town Gas Coal-based syngas provided by municipalities for heating and lighting, primarily in the nineteenth century.

Underground Mining The extraction of coal or its products from between enclosing rock strata by underground mining methods, such as room and pillar, longwall, and shortwall, or through *in situ* gasification.

Undiscovered Resources Unspecified bodies of coal surmised to exist on the basis of broad geologic knowledge and theory. Undiscovered resources include beds of bituminous coal and anthracite 14 inches or more thick and beds of subbituminous coal and lignite 30 inches or more thick that are presumed to occur in unmapped and unexplored areas to depths of 6,000 feet. The speculative and hypothetical resource categories comprise undiscovered resources.

Virgin Coal Coal that has not been accessed by mining. See also *accessed.*

Water–Gas Shift Reaction An inorganic chemical reaction in which water and carbon monoxide react to form carbon dioxide and hydrogen (water splitting): carbon monoxide (CO) + water (H_2O) + catalyst and heat results in carbon dioxide (CO_2) + hydrogen (H_2).

Wet Bottom Boiler/Furnace A pulverized fuel fired furnace in which the ash particles are deposited and retained on the floor thereof and molten ash is removed by tapping either continuously or intermittently.

Index

Note: Page numbers with "f" denote figures; "t" tables.

D

E

F

U

V

W